T0214087

Communications in Computer and Information Science   **1292**

Commenced Publication in 2007
Founding and Former Series Editors:
Simone Diniz Junqueira Barbosa, Phoebe Chen, Alfredo Cuzzocrea,
Xiaoyong Du, Orhun Kara, Ting Liu, Krishna M. Sivalingam,
Dominik Ślęzak, Takashi Washio, Xiaokang Yang, and Junsong Yuan

Andrey Filchenkov · Janne Kauttonen ·
Lidia Pivovarova (Eds.)

Artificial Intelligence and Natural Language

9th Conference, AINL 2020
Helsinki, Finland, October 7–9, 2020
Proceedings

 Springer

Editors
Andrey Filchenkov
ITMO University
St. Petersburg, Russia

Janne Kauttonen (iD)
Haaga-Helia University of Applied Sciences
Helsinki, Finland

Lidia Pivovarova
University of Helsinki
Helsinki, Finland

ISSN 1865-0929   ISSN 1865-0937   (electronic)
Communications in Computer and Information Science
ISBN 978-3-030-59081-9   ISBN 978-3-030-59082-6   (eBook)
https://doi.org/10.1007/978-3-030-59082-6

This Springer imprint is published by the registered company Springer Nature Switzerland AG
The registered company address is: Gewerbestrasse 11, 6330 Cham, Switzerland

Preface

The 9th Conference on Artificial Intelligence and Natural Language Conference (AINL 2020), held during October 7–9, 2020, in Helsinki, Finland, was organized by NLP Seminar (Russia), ITMO University (Russia), and Haaga-Helia University of Applied Sciences (Finland). It aimed to bring together experts in the areas of text mining, speech technologies, dialogue systems, information retrieval, machine learning, artificial intelligence, and robotics to create a platform for sharing experience, extending contacts, and searching for possible collaboration. Since 2012, the AINL conference series has established a strong foothold in the Baltic region and Russia with a strong focus on research and building resources for local languages.

The reviewing process was challenging. Overall, 36 papers were sent to the conference and only 14 (3 short and 11 full) were accepted (an acceptance rate of 39%). 55 researchers from different domains and areas were engaged in the double-blind review process. Each paper received at least three reviews and, almost half of the papers received four reviews.

The selected papers were presented at the conference, covering a wide range of topics, including sentiment analysis, lexical resources, document retrieval, aspect extraction, dialog modeling, text summarization, text generation, explainable artificial intelligence, community detection, and learning to rank in graphs. Most of the presented papers were devoted to processing textual data. In addition, the conference program included industry talks, workshops, tutorials, and a poster session.

The conference also hosted a Russian-Chinese machine translation shared task. The paper describing the shared task is included into this volume after a non-blind review by a volume editor.

Many thanks to everybody who submitted papers, who made wonderful talks, and to those who participated without publication. We are indebted to our Program Committee members for their insightful reviews.

We are grateful to our organization team: Anastasia Bodrova, Irina Krylova, Dmitry Ustalov, Lili Aunimo, and Haaga-Helia (all from the University of Applied Sciences) for support[1].

October 2020

Andrey Filchenkov
Janne Kauttonen
Lidia Pivovarova

---

[1] This year's preparation and execution of the conference have been heavily influenced by the COVID-19 pandemic. Originally planned to take place at Helsinki in Finland, AINL 2020 was held as a fully digital conference during October 7–9.

Organization

Program Committee

Mikhail Alexandrov	Autonomous University of Barcelona, Spain
Ilseyar Alimova	Kazan Federal University, Russia
Lili Aunimo	Haaga-Helia University of Applied Sciences, Finland
Amir Bakarov	National Research University Higher School of Economics, Russia
Alberto Barrón-Cedeño	University of Bologna, Italy
Erind Bedalli	University of Elbasan, Albania
Anton Belyy	Johns Hopkins University, USA
Siddhartha Bhattacharyya	RCC Institute of Information Technology, Kolkata, India
Elena Bolshakova	Moscow State Lomonosov University, Russia
Pavel Braslavski	Ural Federal University, Russia
Maxim Buzdalov	ITMO University, Russia
John Cardiff	ITT Dublin, Ireland
Boris Dobrov	Moscow State University, Russia
Ekaterina Enikeeva	Saint Petersburg State University, Russia
Llorenç Escoter	Google, Switzerland
Vera Evdokimova	Saint Petersburg State University, Russia
Andrey Filchenkov	ITMO University, Russia
Simon Hengchen	University of Gothenburg, Sweden
Dmitry Ignatov	National Research University Higher School of Economics, Russia
Vladimir Ivanov	Innopolis University, Russia
Janne Kauttonen	Haaga-Helia University of Applied Sciences, Finland
Denis Kirjanov	National Research University Higher School of Economics, Russia
Daniil Kocharov	Saint Petersburg State University, Russia
Mikhail Korobov	ScrapingHub Inc., Russia
Evgeny Kotelnikov	Vyatka State University, Russia
Dmitry Kravchenko	Ben-Gurion University of the Negev, Israel
Andrey Kutuzov	University of Oslo, Norway
Elizaveta Kuzmenko	University of Trento, Italy
Natalia Loukachevitch	Research Computing Center of Moscow State University, Russia
Alexey Malafeev	National Research University Higher School of Economics, Russia
Valentin Malykh	ISA RAS, Russia
Vladislav Maraev	University of Gothenburg, Sweden

George Mikros	National and Kapodistrian University of Athens, Greece
Tristan Miller	Austrian Research Institute for Artificial Intelligence, Austria
Kirill Nikolaev	National Research University Higher School of Economics, Russia
Georgios Petasis	NCSR Demokritos, Greece
Jakub Piskorski	Joint Research Centre of the European Commission, Poland
Lidia Pivovarova	University of Helsinki, Finland
Vladimir Pleshko	RCO, Russia
Paolo Rosso	Universitat Politècnica de València, Spain
Yuliya Rubtsova	IIS SB RAS, Russia
Eugen Ruppert	Universität Hamburg, base.camp, Germany
Andrey Savchenko	National Research University Higher School of Economics, Russia
Christin Seifert	University of Twente, The Netherlands
Anastasia Shimorina	University of Lorraine, LORIA, France
Irina Temnikova	Qatar Computing Research Institute, Qatar
Elena Tutubalina	Kazan Federal University, Russia
Dmitry Ustalov	Yandex, Russia
Elior Vila	University of Elbasan, Albania
Nikolai Zolotykh	University of Nizhni Novgorod, Russia

Additional Reviewers

Bolotova, Valeriya
Hernandez Farias, Delia Irazu

Contents

PolSentiLex: Sentiment Detection in Socio-Political Discussions on Russian Social Media

Olessia Koltsova[1] [ID], Svetlana Alexeeva[2] [ID], Sergei Pashakhin[1($\boxtimes$)] [ID], and Sergei Koltsov[1] [ID]

[1] Laboratory for Social and Cognitive Informatics, National Research University Higher School of Economics, Room 216, 55/2 Sedova Street, 190008 St. Petersburg, Russia
{ekoltsova,spashahin,skoltsov}@hse.ru

[2] Laboratory for Cognitive Studies, St. Petersburg State University, St. Petersburg, Russia
mail@s-alexeeva.ru

Abstract. We present a freely available Russian language sentiment lexicon PolSentiLex designed to detect sentiment in user-generated content related to social and political issues. The lexicon was generated from a database of posts and comments of the top 2,000 LiveJournal bloggers posted during one year (~1.5 million posts and 20 million comments). Following a topic modeling approach, we extracted 85,898 documents that were used to retrieve domain-specific terms. This term list was then merged with several external sources. Together, they formed a lexicon (16,399 units) marked-up using a crowdsourcing strategy. A sample of Russian native speakers (n = 105) was asked to assess words' sentiment given the context of their use (randomly paired) as well as the prevailing sentiment of the respective texts. In total, we received 59,208 complete annotations for both texts and words. Several versions of the marked-up lexicon were experimented with, and the final version was tested for quality against the only other freely available Russian language lexicon and against three machine learning algorithms. All experiments were run on two different collections. They have shown that, in terms of F_{macro}, lexicon-based approaches outperform machine learning by 11%, and our lexicon outperforms the alternative one by 11% on the first collection, and by 7% on the negative scale of the second collection while showing similar quality on the positive scale and being three times smaller. Our lexicon also outperforms or is similar to the best existing sentiment analysis results for other types of Russian-language texts.

Keywords: Social media · Socio-political domain · Sentiment analysis · Russian language · Lexicon-based approach

This work is an output of a research project implemented as part of the Basic Research Program at the National Research University Higher School of Economics (HSE University).

© Springer Nature Switzerland AG 2020
A. Filchenkov et al. (Eds.): AINL 2020, CCIS 1292, pp. 1–16, 2020.
https://doi.org/10.1007/978-3-030-59082-6_1

1 Introduction

Tools and resources for sentiment analysis (SA) in the Russian language are often created with a focus on consumer reviews and remain mostly underdeveloped for other types of communication taking place on social media. Increasingly crucial in public life, social media have become valuable for social scientists studying sentiment in online political discussions or interested in predicting public reaction to events with online data. However, such studies face a lack of SA resources, as they are usually language- and domain-specific, or demand expertise in machine learning and feature engineering from their users. Additionally, for the Russian language, the ever obtained quality in various SA tasks is modest even when the most advanced machine learning methods are applied. Our work seeks to overcome these obstacles.

In this article, we make the following contributions:

- We propose *PolSentiLex*, a freely available (https://linis-crowd.org/) and easy to use lexicon for sentiment analysis of social media texts on social and political issues in the Russian language;
- We employ several heuristics, including topic modeling, that allow obtaining the maximum number of domain-specific sentiment words;
- The proposed lexicon—*PolSentiLex*—outperforms the only other freely available Russian-language lexicon that was released after the start of our project as a general-purpose lexicon;
- We demonstrate that, for our case, lexicon-based approaches significantly outperform machine learning, especially on the collection that was not used for lexicon generation.

The paper is organized as follows. The next section presents a review of the literature on sentiment analysis in the Russian language, social media, and socio-political domain. In Sects. 3 and 4, we describe the process of PolSentiLex development and the procedure of its quality assessment, respectively. Section 5 contains the results. We conclude with suggestions for sentiment analysis of socio-political messages from Russian language social media in Sect. 6.

2 Related Work

Lay texts of the socio-political domain present larger difficulties for sentiment classification than consumer product reviews, which is explained by a more subtle expression of sentiment in the former, including the presence of sarcasm and ironies [33]. Usually, socio-political issues are described with larger vocabularies than consumer reviews, and contain a complex mixture of factual reporting and subjective opinions [19]. Moreover, as pointed out by Eisenstein [11], the language on social media where such texts are created significantly differs from the "norm," containing more non-standard spelling and punctuation, vocabulary, and syntax. Additionally, language use varies both within and across social media platforms. Following in line with Androutsopoulos [1] and Darling et al.

[10] Eisenstein argues that, for instance, Twitter language varies a lot in "the degree of standardness." Such domain- and medium-specific features have to be accounted for when applying both lexicon or machine learning (ML) approaches to SA [23,26].

2.1   Sentiment Analysis in the Russian Langauge

The state of sentiment analysis (SA) in the Russian language is reviewed in [30], and the review of existing Russian-language lexicons is available in Kotelnikov et al. [17]. The latter work tests the performance of 10 versions of 8 lexicons on the task of consumer reviews using SVM algorithm. It is indicative that not only the task is related to consumer reviews, but also 6 out of 8 lexicons, including the authors' lexicon, were created specifically for review classification [2,5,7,16,24,32]. The two that were not, are RuSentiLex [8] and the early version of PolSentiLex described in the paper by the title of the website where it was published (Linis-Crowd). All lexicons lose the race to the full dictionary of the corpus, the second place being taken by the united lexicon, and the third by ProductSentiRus [7].

Similar focus on reviews can be seen in sentiment analysis tracks hosted by Russian Information Retrieval Seminar (ROMIP) in 2012–2016. Apart from consumer reviews, these tracks have also included tasks on social media texts (blogs and Twitter) and news items [6,9,20,22]. However, except for the latter, during all events, participants have been offered to detect sentiment in opinions about services, products, and organizations. As expected, most winners follow machine learning approaches, with SVM on average being the most accurate algorithm. Nonetheless, most of those solutions rely on manually created dictionaries of sentiment-bearing words or features engineered based on such resources. Lexicon-based approaches, on the other hand, show worst and rarely comparable results with only one exception (see below).

ROMIP participants and organizers note that the performance of solutions is highly dependent on the number of sentiment classes, data sources, and task domains. Blogs and tweets as well as texts from the socio-political domain in general are considered the most difficult. While general social media texts, compared to consumer reviews, are reported to be more diverse in sentiment expression, less structured and less grammaticaly sound, the difficulty of socio-political texts is attributed to the greater variety of subtopics, compared to other domains. Interestingly, the only occasion where a lexicon-based approach outperformed machine learning was during the task of detecting sentiment in news items [18]. The winner solution used an extensive dictionary of opinion words and expressions obtained manually as well as with text mining techniques. The system estimated sentiment following several rules regarding sentiment intensifiers, negation, and following opinion words. Although significantly better than the baseline, this solution showed $F_{macro} = 0.62$. All this points at the need for the development of SA resources for Russian-language general interest social media texts and for socio-political texts, including professional news and lay messages.

This lack of generalist SA resources for the Russian language was addressed by Loukachevitch and Levchik [21] to create a general sentiment lexicon named *RuSentiLex*. It is the successor of the very first publicly available Russian sentiment lexicon, which had no polarity scores and was developed by the same researchers [8]. RuSentiLex was constructed in a semi-automatic way from several sources: (1) words from domain-specific lexicons matched with Russian language thesaurus; (2) words and phrases extracted following special rules from a large corpus of news articles; (3) slang and curse words extracted from Twitter with a supervised model of sentiment word extraction. The lexicon consists of 16,057 n-grams, with 63.7% of them being negative, 23.6% – positive, 10.9% – neutral, and with 1.8% having context-depended sentiment polarity.

To the best of our knowledge, the only work that tested the quality of RuSentiLex is the one by Kotelnikov et al. [17]; who, as it was noted, focused on consumer reviews. Also, some ROMIP participants successfully used RuSenitLex for feature engineering which allowed them to beat either the baseline or some other competitors [21]. As RuSentiLex is, to the best of our knowledge, the only freely available Russian-language lexicon not aimed at review classification, in this work we use it as our baseline. We test both RuSentiLex and PolSentiLex using a rule-based algorithm designed for them and further compare them to several algorithms of machine learning on two different datasets.

3   PolSentiLex

In this section, we briefly review the process of PolSentiLex construction. It closely follows the procedure adopted for the early version of our lexicon [14] that had produced only a very modest quality; all differences of this version are accounted for further below.

3.1   LiveJournal Collection of Social and Political Posts

Given our primary goal to develop a lexicon for SA of messages from politicized social media, we began with extracting domain-specific words. We constructed two text collections that contain posts and comments from top 2,000 accounts in the Russian language section of LiveJournal for the period of one year (March 2013–March 2014). At the time of data collection, LiveJournal was still an active and a highly politicized blog platform in Russia. However, our previous research based on this data indicates that only about one-third of these documents can be classified as political or social, which posed a problem of retrieving relevant texts out of approximately 1.5 million posts and 0.9 million comments.

To solve this problem, we performed Latent Dirichlet Allocation (LDA) topic modeling with Gibbs sampling [12] on the collections of posts and comments separately. The early version of our lexicon [14] used only posts as presumably more informative in terms of topic and sentiment vocabulary. In this work we introduce the merged lexicon which also uses the collection of comments that were added due to the obviously insufficient quality of the first version. To overcome poor performance of topic modeling on short texts, all comments related to the same post were merged. Based on experiments performed with similar data and described in [4, 25], we modeled both collections with 300 topics.

Next, each topic was manually labelled by three annotators from our team based on reading of a maximum of 200 top terms and 100 top texts (ranked by their probability in a topic). They identified 104 and 88 topics from social or political domains in each of the two collections of posts and comments, respectively. Additionally, nine and 20 topics composed mostly from general emotional words were identified in the collections of posts and comments, respectively. We considered a topic to be relevant to our chosen domains if at least two annotators agreed on that.

Finally, from each of the relevant topics we retrieved the most relevant texts based on the values from the topic-document matrix Φ. For post collection, the threshold of relevance was set to (>0.1) which produced a subsample of 70,710 blog posts, and for the comment collection it had to be set lower (>0.001), which yielded a smaller subsample of 15,188 merged comment texts.

3.2   Selection of Potentially Sentiment-Bearing Words

We created the core of our proto-lexicon using the list of the top 200 words from all social and political topics. Then, we extended our list of potentially sentiment-bearing terms with the words from several external sources described in detail in [14].

Next, we intersected the listed sources and retrieved only the terms that occurred in at least two of them. The dictionary that resulted from the work with post collection contained 9,539 units, and the one resulting from the comment collection consisted of 6,860 units.

3.3   Data Mark Up

One of our main ideas in the lexicon construction was that words, even those retrieved from general dictionaries, might have specific sentiment polarity or strength when occurring in social or political texts. Therefore, we chose to mark up the selected words providing the annotators with the context in which the words occurred; simultaneously, it allowed us to combine word mark up with text mark up.

As both negative and positive sentiment in texts can produce a social effect so far as it is perceived as such by society members, polarity scores for our lexicon and document collections were obtained from lay native speakers. Therefore, our assessors were not supposed to imitate experts; instead, we defined their contribution similar to that of respondents in an opinion poll: no "wrong" answers were possible. In total, 87 people took part in the assessment of posts, and 18 individuals from our team participated in the mark up of comments. Volunteers for post annotation were recruited via social media. All of them participated in instruction sessions and training annotation, where all of them were offered the same texts, after which some coders were discarded.

A special website (https://linis-crowd.org/) was designed for the mark-up which asked participants to assess words' sentiment as expressed in the texts where they occurred, as well as the prevailing sentiment of texts themselves, with a five-point scale, from -2 (strongly negative) to $+2$ (strongly positive). For each word, the system randomly selected three different texts relevant to politics or public affairs. Since some texts were not unique for each word, the texts, too, received multiple scores.

Each word was coded three times but not necessary in three different texts; some words from our proto-dictionary did not occur in our collections and were excluded. Also, since a word-text pair was randomly selected, several pairs were coded more than three times. As a result, at our first stage we received 32,437 annotation points both for posts and words: 7,546 unique words were annotated three times each, and 19,831 unique posts received one or more marks. At our second stage we repeated the entire procedure on the comment collection and obtained 26,851 annotation points for both merged comment texts and words, with 6,860 unique words receiving three marks each and with 15,188 unique comment texts received at least one mark. Intercoder agreement calculated among three random grades for each of the words is 0.553 in terms of Krippendorf's alpha. In the resulting lexicon, all grades of each word were averaged and rounded.

3.4   The Three Versions of PolSentiLex

In the course of all the experiments we tested three versions of our lexicon. The first version (further, *post version*) included 2,793 non-neutral words derived from the collection of social and political blog posts. The rest of 7,546 annotated words were found to carry no sentiment. This lexicon produced the quality of 0.44 in terms of F_{macro}, reported in [14] which is why further experiments were carried out. The next version (further, *comment version*) included 2,664 non-neutral words derived from the collection of merged comment texts. The experiments with it (not reported) produced no increase in quality. Eventually, we combined both lexicons into the final version (further, *combined version*); since some words occurred in both the post and the comment versions, their scores were averaged.

Table 1 shows the distributions of scores over words in the three versions of PolSentiLex. Obviously, although negatively assessed words prevail, positive words are also present. At the same time, very few highly emotional words are observed.

4   PolSentiLex Quality Assessment

In this section, we describe experiments in which we evaluate the quality of PolSentiLex against RuSentiLex [21]. These two lexicons are used as feature sets in three machine learning algorithms and in a dictionary-based technique, and tested on two datasets.

Table 1. Distribution of mean scores over words in the post, comment and combined versions of PolSentiLex

Mean score (rounded)	N words	Share of words, %
Post lexicon		
−2	225	3
−1	1,666	22
0	4,753	63.4
1	853	11
2	49	0.6
In total	7,546	100
Not neutral (in total)	2,793	37
Comment lexicon		
−2	173	2.5
−1	1,882	27.3
0	4,196	61
1	596	9
2	13	0.2
In total	6,860	100
Not neutral (in total)	2,664	39
Combined lexicon		
−2	252	2
−1	2,612	24
0	6,738	63.8
1	1,031	10
2	29	0.2
In total	10,662	100
Not neutral (in total)	3,924	37

4.1 Datasets

Both corpora used for quality assessment of our lexicon are comprised of socio-political texts from social media: one is a subsample of posts used to create our lexicon (see Sects. 3.1 and 3.3, further *LiveJournal posts*) and the other is an independent corpus (further *Ethnicity collection*).

The Ethnicity collection was sampled from all possible Russian social media and blogs for the period from January 2014 to December 2015 that contained texts with an ethnonym. Ethnonyms—linguistic constructions used to nominate ethnic groups—were derived from a predefined list of 4,063 words and bigrams covering 115 ethnic groups living in Russia. Based on this list, the sample was provided by a commercial company that collects the Russian language social media content. This collection was chosen for this project out of all the Russian-language collections known to us as the most relevant to the task of testing a socio-political lexicon (discussions of ethnicity are usually not only highly politicized, but also very heated). Furthermore, it was not used in the development of either PolSentiLex or RuSentiLex. The well-known RuSentiment dataset [29] that was made available a little later is not focused on political texts.

From our Ethnicity collection, 14,998 messages were selected so as to represent all ethnic groups and assessed by three independent coders each; of them 12,256 were left after filtering near-duplicates. The coders were asked to answer a number of questions about the texts, including two most important for this study: (a) how strongly a general negative sentiment is expressed in the text, if any? (no/weak/strong); (b) how strongly a general positive sentiment is expressed in the text, if any? (no/weak/strong). In this mark-up, we used two independent scales for positive and negative sentiment instead of the integral sentiment scale used for LiveJournal collection (see Sects. 3.1 and 3.3) as it corresponded better to the purpose for which Ethnicity collection was constructed.

While LJ collection was marked-up in parallel with word mark-up, as explained in Sect. 3.3, with the same set of annotators, marking-up of Ethnicity collection was a separate task, but it followed a similar procedure, with 27 student volunteers being specially trained for that. Intercoder agreement, as expressed with Krippendorf's alpha is, on LJ collection: 0.541 for a five-class task, on Ethnicity collection: 0.547 on the negative scale, and 0.404 on the positive scale, both being three-class tasks. This level of agreement is quite common in sentiment analysis [3]. Texts that received fewer than two marks were excluded.

Grades for the Ethnicity collection could vary from -2 to 0 for the negative scale and from 0 to $+2$ for the positive scale, and all these categories turned out to be reasonably populated. However, the LiveJournal collection which had been marked-up on a unified positive-negative scale, turned out to have very few texts with the extreme values $+2$ or -2 (about 6%). Therefore, we collapsed the five-point $(-2, -1, 0, 1, 2)$ scale into a three-point scale $(-1, 0, 1)$ where $-2 = -1$ and $2 = 1$. As a result, we formed three-class classification tasks that thus became easier to compare. Final polarity scores for all texts were calculated as the mean values of individual human grades.

Table 2 shows the distribution of scores over texts. Most texts are marked as neutral or negative, with fewer positive marks. For the LiveJournal collection, the positive to negative class proportion is 1:5.8, and for the Ethnicity collection it is 1:3.2. The same unbalanced class structure in political blogs is also pointed at by Hsueh et al. [13].

Before testing, both LiveJournal and Ethnicity collections were preprocessed: for the ML approach, we cleaned each collection from non-letter symbols and lemmatized each word with Pymorphy2 [15]; and for the lexicon approach, we used lemmatized documents with punctuation intact.

We performed multiple comparisons of PolSentiLex and RuSentiLex used as feature sets in one rule-based approach and in three ML algorithms. Based on preliminary experiments, we chose the version of our lexicon that performed better or not worse than the other versions. Predictably, it turned out to be the

Table 2. Distribution of mean scores over text in the LiveJournal posts and Ethnicity collection

Mean score (rounded)	N texts	Share of texts, %
LiveJournal posts, integral score		
−1	2,104	33
0	3,940	61
1	360	6
In total	6,404	100
Not neutral (in total)	2,464	38
Ethnicity collection		
Negative scale		
−2	1,126	9
−1	4,181	33.2
0	7,272	57.8
In total	12,579	100
Not neutral (in total)	5,307	42
Positive scale		
0	10,882	86.6
1	1,436	11.4
2	261	2
In total	12,579	100
Not neutral (in total)	1,652	13

combined version (see Sect. 3.4) of PolSentiLex that comprised of 3,924 terms. As for RuSentiLex, we used all not-neutral context-independent unigrams that counted to 11,756 units in total because unigrams were reported to be the most useful features for sentiment classification [27].

In total, we performed 24 tests: three tasks (negative and positive sentiment prediction for the Ethnicity collection and overall polarity prediction for the LiveJournal collection) were performed with the two lexicons, each of which was used with four approaches (thee ML algorithms and one rule-based). The total number of runs, including all parameter optimization and cross-validation iterations, was 390.

For the ML approach, we chose the three most popular algorithms for SA, namely support vector machine (SVM) with linear basis function kernel, Gaussian Naïve Bayes (NB), and k-nearest neighbors (KNN) classifier. For training, we used a document-term matrix with the presence of a term from a lexicon in documents. We used a random 75% data sample for training and validation, and the rest 25% for testing (held-out data). First, following the grid-search strategy with 2-fold cross-validation, we identified the best parameters for SVM and KNN on training data. For SVM, we explored hyper-parameter C in the range

[0.0001, 0.001, 0.01, 0.1, 1, 100, 1000, 10000, 100000, 1000000] and identified that the algorithm performed best with C = 0.0001. For KNN, we varied the number of neighbors from 1 to 40, and in almost all tests, the best performance was achieved with k = 1. The two exceptions were the Ethnicity collection with PolSentiLex as a feature set on a positive scale (k = 3) and the Ethnicity collection with PolSentiLex as a feature set on a negative scale (k = 4). Then, using the obtained parameters, we trained each classifier with 10-fold cross-validation. Finally, to obtain an unbiased evaluation, we applied a classifier with the highest F_{macro} on validation data to holdout datasets. To train classifiers, we used the Scikit-learn Python library [28].

For lexicon approach, we used SentiStrength rule-based algorithm [31]. We chose SentiStrength because its implementation is freely available, and it was designed specifically for the social web texts. To classify a document, SentiStrength, firstly, searches the text for and scores terms from a sentiment dictionary defined by a user, correcting their scores for the presence of booster and negation words. It then applies one of several approaches to estimate sentence- and text-level sentiment on positive and negative scales separately. Based on the preliminary experiments, we chose the approach that showed the best results where the sentiment of a sentence equals to the sentiment of its strongest term, and the sentiment of a text equals the strongest sentence sentiment.

To accurately assess the quality of SentiStrength prediction, we had to transform its text sentiment scores so that they become comparable to the classes from the human mark-up. Because of booster words, SentiStrength sentiment score for a text could go beyond ±2. Therefore, to align the scales of predicted sentiment scores and the true assessors' scores for Ethnicity collection (0,+1,+2 & 0,−1,−2), we considered all texts with the SentiStrength score above +2 to belong to the class "+2" (highly positive), and all texts with the SentiStrength scores below −2 to belong to class "−2" (highly negative).

As LiveJournal collection texts had been marked up using a single sentiment scale (from −2 to 2), we applied the following steps to transform two separate SentiStrength scores into a single score and to compare them to the respective human scores. (1) We calculated the mean of the two SentiStrength scores (positive and negative) and thus obtained the integral predicted score for each text, PS. (2) We calculated the difference between PS and the true score, TS, taken as the non-rounded assessors' sentiment score for the same text. (3) As both TS, PS and their difference were likely to be non-integer, to determine whether the true classes were correctly predicted, we used the following logical expressions: (3a) If $|PS - TS| < 0.5$, then $PS = TS$, i.e. the true class is correctly predicted. (3b) If $0.5 \leqslant |PS - TS| < 1.5$ then $|PS - TS| = 1$, i.e, the classification error is ±1 class. (3c) If $|PS - TS| \geqslant 1.5$ then $|PS - TS| = 2$, i.e. the classification error is ±2 classes.

To evaluate the performance of all our sentiment analysis approaches, we used standard metrics for classifier performance: the F_{macro} measure (reported in Fig. 1), precision (reported in Fig. 2), recall (reported in Fig. 3) and accuracy (reported in Fig. 4).

5   Results

The most important results are presented in Figs. 1, 2, 3 and 4 and Table 3. First, our best solutions (KNN with PolSentiLex for accuracy on the positive scale of Ethnicity collection and SentiStrength with PolSentiLex for all other tasks and quality metrics) significantly exceed the random baseline, accounting for the class imbalance. Thus, in terms of accuracy the gain over the random baseline is 14–51% which is very good for the Russian language. For instance, similar ROMIP tasks on three-class sentiment classification of news and on consumer blog posts demonstrated the gain of 5–49% on average [14]. Moreover, our best solution—PolSentiLex with SentiStrength—has also improved a lot as compared to our previous result [14]. On the LiveJournal collection, which was used for testing our lexicon last time, the improvement has been 14% and 13% in terms of precision and recall, respectively. Comparing performance of our new lexicon on the Ethnicity collection, from which it was not derived, and the performance of our old lexicon on LiveJournal collection, from which we did derive it, the new lexicon is still 3–12% better in both precision and recall on the negative scale and in recall on the positive scale, although it is slightly worse in precision on the negative scale.

Table 3. Advantage of PolSentiLex over RuSentiLex using SentiStrength

	F_{macro}	Precision	Recall	Accuracy
LiveJournal	11%	9%	13%	10%
Ethnicity - negative	8%	4%	11%	0%
Ethnicity - positive	0%	0%	0%	7%

Second, interestingly, rule-based approaches with any lexicons are visibly better than any ML approaches on all datasets and across all metrics, with one exception addressed further below. Thus, in terms of Fmacro measure, the lexicon approaches perform, on average, 11% better than the ML approaches, which is a huge difference. It might be attributed to a non-optimal parameter choice in our ML solutions, however, the two best ML approaches (KNN and SVM) produce the gain over baseline comparable to that in ROMIP tracks [14].

A look at the exception—namely, the KNN method with PolSentiLex for the positive scale in terms of general accuracy—reveals a curious result: KNN solution does not exceed lexicon-based solutions either in precision or recall. A closer examination of the relevant confusion matrices shows that KNN ascribes almost all texts to the neutral class producing exceptionally low precision and recall for the classes +1 and +2, i.e., it often fails to detect positive sentiment. However, as the non-positive class in the Ethnicity collection is by far larger than the other two and constitutes 86%, a fair ability of KNN to detect this class contributes a lot to the overall accuracy (84%). This result is suboptimal for social scientists who usually aim to detect texts containing non-neutral emotions, which is what rule-based approaches perform better, albeit at the expense of neutral class detection. We can assume that lexicon-based approaches might be better for social science tasks including ours, which is consistent with Thelwall's conclusions [31] and with the ROMIP results reported in Sect. 2. This means that, unlike consumer reviews, politicized social media texts are more diverse, less structured and are harder to divide into classes, which might make manually selected words more reliable class indicators than features engineered with ML.

Finally, the most important comparison is that between the two tested lexicons (see Table 3). The fact that our lexicon outperforms RuSentiLex on the LiveJournal collection is predictable since this collection was used as a source for our lexicon, but not for its competitor. A more interesting observation is that on the two other tasks PolSentiLex is also either better or not worse than RuSentiLex in terms of all aggregated metrics. This deserves mentioning given that our lexicon is only 33% the size of RuSentiLex (3,924 words against 11,756). However, to give a fair treatment to RuSentiLex, we should look into respective confusion matrices and the distribution of quality metric values over classes.

On the positive scale, PolSentiLex, on average, has no advantage over RuSentiLex in precision, while its recall is higher for the neutral class and lower for class (+1). The overall recall of PolSentiLex on both sentiment-bearing classes is about 5% lower than that of RuSentiLex, even though the former has a slight advantage over the latter in class (+2). Since, as it has been mentioned, prediction of sentiment-bearing classes is a priority for social science tasks, PolSentiLex cannot be considered clearly better in predicting positive sentiment than RuSentiLex, despite the better overall accuracy of the former.

On the negative scale, with PolSentiLex having some advantage in precision, it yields a visibly smaller recall for class (−1). However, here the main confusion of PolSentiLex is not that severe, being between classes (−1) and (−2). It means that PolSentiLex is not prone for losing negative texts; instead, it tends to overestimate negative sentiment by classifying some moderately negative texts as highly negative. At the same time, PolSentiLex is much better in both precision and recall on class (−2), and its overall accuracy on the two sentiment-bearing classes is marginally higher than that of RuSentiLex. We can conclude that the two lexicons are similar in quality, especially in precision, and that there is a trade-off between overall accuracy and the ability to detect sentiment-bearing classes.

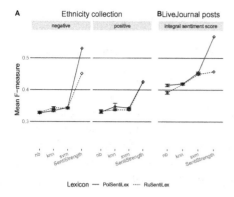

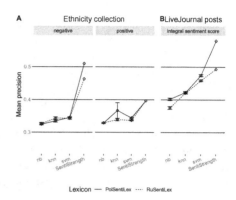

Fig. 1. F_{macro} results: a) the Ethnicity collection, b) the collection of LiveJournal posts.

Fig. 2. Precision results: a) the Ethnicity collection, b) the collection of LiveJournal posts.

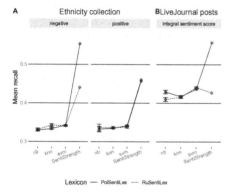

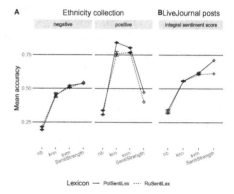

Fig. 3. Recall results: a) the Ethnicity collection, b) the collection of LiveJournal posts.

Fig. 4. Accuracy results: a) the Ethnicity collection, b) the collection of LiveJournal posts.

6   Conclusion

The results suggest that in sentiment analysis of socio-political messages from Russian-language social media, given the available resources, a social scientist will be better off with using a rule-based method, such as provided by SentiStrength package, with either PolSentiLex or RuSentiLex lexicons. With any of them, a user will not only get a visibly higher quality, but also lower computational complexity and a much more user-friendly and intuitive method. While PolSentiLex shows lower recall for moderate classes (moderately positive and moderately negative texts), it is either better or not worse than RuSentilex in detection of all other classes, according to all metrics, including those aggregated over all classes. Since PolSentiLex is also much smaller than RuSentiLex, it might be considered an optimal choice for the time being, although further improvements are needed.

One of the directions for improvement is to merge the lexicons while giving priority to PolSentiLex mark-up and re-evaluating the polarity of the remaining RuSentiLex terms in a socio-political context. Another improvement might be gained by adding bigrams typical for socio-political texts, starting with those that contain sentiment words. Next, text mark-up may also be used as a source for lexicon enrichment: thus, assessors may be asked to mark text fragments that were most helpful to form their opinion on the text. Finally, both lexicon might be used as features in more advanced machine learning algorithms such as neural networks, along with distributed word representations.

References

1. Androutsopoulos, J.: Language change and digital media: a review of conceptions and evidence. In: Standard Languages and Language Standards in a Changing Europe, pp. 145–160. Novus, Oslo (2011)
2. Blinov, P.D., Klekovkina, M.V., Kotelnikov, E.V., Pestov, O.A.: Research of lexical approach and machine learning methods for sentiment analysis. In: Computational Linguistics and Intellectual Technologies: Papers from the Annual International Conference "Dialogue-2013", vol. 2, pp. 51–61. RGGU, Moscow (2013). http://www.dialog-21.ru/media/1226/blinovpd.pdf
3. Bobicev, V., Sokolova, M.: Inter-annotator agreement in sentiment analysis: machine learning perspective. In: Proceedings of the International Conference on Recent Advances in Natural Language Processing, RANLP 2017, pp. 97–102. INCOMA Ltd., Varna, September 2017. https://doi.org/10.26615/978-954-452-049-6_015
4. Bodrunova, S., Koltsov, S., Koltsova, O., Nikolenko, S., Shimorina, A.: Interval semi-supervised LDA: classifying needles in a haystack. In: Castro, F., Gelbukh, A., González, M. (eds.) MICAI 2013. LNCS (LNAI), vol. 8265, pp. 265–274. Springer, Heidelberg (2013). https://doi.org/10.1007/978-3-642-45114-0_21
5. Chen, Y., Skiena, S.: Building sentiment lexicons for all major languages. In: Proceedings of the 52nd Annual Meeting of the Association for Computational Linguistics (Volume 2: Short Papers), pp. 383–389. Association for Computational Linguistics, Baltimore (2014). https://doi.org/10.3115/v1/P14-2063, http://aclweb.org/anthology/P14-2063
6. Chetviorkin, I., Braslavski, P., Loukachevitch, N.: Sentiment analysis track at ROMIP 2011. In: Computational Linguistics and Intellectual Technologies: Papers from the Annual International Conference "Dialogue", vol. 2, pp. 1–14 (2012) (2012)
7. Chetviorkin, I., Loukachevitch, N.: Extraction of Russian sentiment lexicon for product meta-domain. In: Proceedings of COLING 2012: Technical Papers, pp. 593–610. The COLING 2012 Organizing Committee, Mumbai (2012). https://www.aclweb.org/anthology/C12-1037
8. Chetviorkin, I., Loukachevitch, N.: Extraction of Russian sentiment lexicon for product meta-domain. In: Proceedings of COLING 2012: Technical Papers, Mumbai, pp. 593–610, December 2012

9. Chetviorkin, I., Loukachevitch, N.: Sentiment analysis track at ROMIP 2012. In: Computational Linguistics and Intellectual Technologies (2013). http://www. dialog-21.ru/digests/dialog2013/materials/pdf/1_ChetverkinII.pdf

10. Darling, W., Paul, M., Song, F.: Unsupervised part-of-speech tagging in noisy and esoteric domains with a syntactic-semantic Bayesian HMM. In: Proceedings of the 13th Conference of the European Chapter of the Association for Computational Linguistics. Association for Computational Linguistics, Avignon (2012)

11. Eisenstein, J.: What to do about bad language on the internet. In: Proceedings of the 2013 Conference of the North American Chapter of the Association for Computational Linguistics: Human Language Technologies, pp. 359–369 (2013)

12. Griffiths, T.L., Steyvers, M.: Finding scientific topics. Proc. Natl. Acad. Sci. **101**(Suppl. 1), 5228–5235 (2004)

13. Hsueh, P.Y., Melville, P., Sindhwani, V.: Data quality from crowdsourcing: a study of annotation selection criteria. In: Proceedings of the NAACL HLT 2009 Workshop on Active Learning for Natural Language Processing, pp. 27–35. Association for Computational Linguistics (2009)

14. Koltsova, O., Alexeeva, S., Koltsov, S.: An opinion word lexicon and a training dataset for Russian sentiment analysis of social media. In: Computational Linguistics and Intellectual Technologies: Proceedings of the International Conference "Dialogue 2016", pp. 277–287. RSUH, Moscow (2016)

15. Korobov, M.: Morphological analyzer and generator for Russian and Ukrainian languages. In: Khachay, M.Y., Konstantinova, N., Panchenko, A., Ignatov, D.I., Labunets, V.G. (eds.) AIST 2015. CCIS, vol. 542, pp. 320–332. Springer, Cham (2015). https://doi.org/10.1007/978-3-319-26123-2_31

16. Kotelnikov, E., Bushmeleva, N., Razova, E., Peskisheva, T., Pletneva, M.: Manually created sentiment lexicons: research and development. In: Computational Linguistics and Intellectual Technologies: Papers from the Annual International Conference "Dialogue-2016", vol. 15, pp. 300–314. RGGU, Moscow (2016). http:// www.dialog-21.ru/media/3402/kotelnikovevetal.pdf

17. Kotelnikov, E., Peskisheva, T., Kotelnikova, A., Razova, E.: A comparative study of publicly available russian sentiment lexicons. In: Ustalov, D., Filchenkov, A., Pivovarova, L., Žižka, J. (eds.) AINL 2018. CCIS, vol. 930, pp. 139–151. Springer, Cham (2018). https://doi.org/10.1007/978-3-030-01204-5_14

18. Kuznetsova, E., Loukachevitch, N., Chetviorkin, I.: Testing rules for a sentiment analysis system. In: Computational Linguistics and Intellectual Technologies: Proceedings of the International Conference "Dialogue 2013", vol. 2, pp. 71–80 (2013). http://www.dialog-21.ru/digests/dialog2013/materials/pdf/KuznetsovaES.pdf

19. Liu, B.: Sentiment Analysis and Opinion Mining. Morgan & Claypool Publishers (2012)

20. Loukachevitch, N., Blinov, P., Kotelnikov, E., Rubtsova, Y., Ivanov, V., Tutubalina, E.: SentiRuEval: testing object-oriented sentiment analysis systems in Russian. In: Computational Linguistics and Intellectual Technologies, p. 13 (2015). http://www.dialog-21.ru/digests/dialog2015/materials/pdf/LoukachevitchNVetal.pdf

21. Loukachevitch, N., Levchik, A.: Creating a general Russian sentiment lexicon. In: Proceedings of Language Resources and Evaluation Conference, LREC-2016, pp. 1171–1176 (2016)

22. Loukachevitch, N., Rubcova, Y.: SentiRuEval-2016: overcoming the time differences and sparsity of data for the reputation analysis problem on Twitter messages [SentiRuEval-2016: preodoleniye vremennykh razlichiy i razrezhennosti dannykh dlya zadachi analiza reputatsii po soobshcheniyam tvittera]. In: Computational Linguistics and Intellectual Technologies: Proceedings of the International Conference "Dialogue 2016", pp. 416–426 (2015)
23. Medhat, W., Hassan, A., Korashy, H.: Sentiment analysis algorithms and applications: a survey. Ain Shams Eng. J. **5**(4), 1093–1113 (2014). https://doi.org/10.1016/j.asej.2014.04.011, http://linkinghub.elsevier.com/retrieve/pii/S2090447914000550
24. Mohammad, S.M., Turney, P.D.: Crowdsourcing a word-emotion association lexicon. Comput. Intell. **29**(3), 436–465 (2013). https://doi.org/10.1111/j.1467-8640.2012.00460.x, http://doi.wiley.com/10.1111/j.1467-8640.2012.00460.x
25. Nikolenko, S., Koltcov, S., Koltsova, O.: Topic modelling for qualitative studies. J. Inf. Sci. **43**(1), 88–102 (2017)
26. Pang, B., Lee, L.: Opinion mining and sentiment analysis. Found. Trends Inf. Retriev. **2**(1–2), 1–135 (2008). https://doi.org/10.1561/1500000001, http://www.nowpublishers.com/article/Details/INR-001
27. Pang, B., Lee, L., Vaithyanathan, S.: Thumbs up?: sentiment classification using machine learning techniques. In: Proceedings of the ACL-2002 Conference on Empirical Methods in Natural Language Processing, vol. 10, pp. 79–86. Association for Computational Linguistics (2002)
28. Pedregosa, F., et al.: Scikit-learn: Machine learning in python. J. Mach. Learn. Res. **12**, 2825–2830 (2011)
29. Rogers, A., Romanov, A., Rumshisky, A., Volkova, S., Gronas, M., Gribov, A.: RuSentiment: an enriched sentiment analysis dataset for social media in Russian. In: Proceedings of the 27th International Conference on Computational Linguistics, pp. 755–763. Association for Computational Linguistics, Santa Fe, August 2018. https://www.aclweb.org/anthology/C18-1064
30. Smetanin, S.: The applications of sentiment analysis for Russian language texts: current challenges and future perspectives. IEEE Access **8**, 110693–110719 (2020). https://doi.org/10.1109/ACCESS.2020.3002215, https://ieeexplore.ieee.org/document/9117010/
31. Thelwall, M., Buckley, K., Paltoglou, G.: Sentiment strength detection for the social web. J. Am. Soc. Inf. Sci. Technol. **63**(1), 163–173 (2012). https://doi.org/10.1002/asi.21662, http://doi.wiley.com/10.1002/asi.21662
32. Tutubalina, E.: Metody izvlecheniya i rezyumirovaniya kriticheskih otzyvov pol'zovatelej o produkcii (Extraction and summarization methods for critical user reviews of a product). Ph.D. thesis, Kazan Federal University, Kazan (2016). https://www.ispras.ru/dcouncil/docs/diss/2016/tutubalina/dissertacija-tutubalina.pdf
33. Zhang, S., Zhang, X., Chan, J., Rosso, P.: Irony detection via sentiment-based transfer learning. Inf. Process. Manage. **56**(5), 1633–1644 (2019). https://doi.org/10.1016/j.ipm.2019.04.006, https://linkinghub.elsevier.com/retrieve/pii/S0306457318307428

Automatic Detection of Hidden Communities in the Texts of Russian Social Network Corpus

Ivan Mamaev[(✉)] [ID] and Olga Mitrofanova [ID]

Saint Petersburg State University,
Universitetskaya emb. 11, Saint Petersburg 199034, Russia
mamaev_96@mail.ru, st079541@student.spbu.ru,
o.mitrofanova@spbu.ru

Abstract. This paper proposes a linguistically-rich approach to hidden community detection which was tested in experiments with the Russian corpus of VKontakte posts. Modern algorithms for hidden community detection are based on graph theory, these procedures leaving out of account the linguistic features of analyzed texts. The authors have developed a new hybrid approach to the detection of hidden communities, combining author-topic modeling and automatic topic labeling. Specific linguistic parameters of Russian posts were revealed for correct language processing. The results justify the use of the algorithm that can be further integrated with already developed graph methods.

Keywords: Hidden communities · Corpus linguistics · Social networks · Author-topic models · Automatic topic labeling

1 Introduction

Contemporary world witnesses a rapid development of IT sphere, and this process covers various areas of everyday life. As a result, the interests of researchers are aimed at social networks, namely to detect communities on the Internet. The problem of detecting communities has recently become the task of network analysis, which allows to understand the structure of social networks. Such parameters as gender, age, geographical location of users, the strength of communication between them, etc. can be analyzed.

Despite significant progress in the field of network analysis, there is a kind of communities which are still hard to detect. Hidden communities are formed by users of social networks with common interests, which are united by latent links. The structure of communities in the real world is not always explicitly expressed, compared to family members or close friends: there are secret organizations, temporary groups, groups of drug addicts, etc. [9] The implicit structure of communities can also be found on social networks basing on users' content similarity.

Our present study is devoted to detection of hidden communities in the Russian segment of the Internet based on the corpus of users' posts in 2018–2020 of VKontakte social network. This website was recognized as one of the most popular online

© Springer Nature Switzerland AG 2020
A. Filchenkov et al. (Eds.): AINL 2020, CCIS 1292, pp. 17–33, 2020.
https://doi.org/10.1007/978-3-030-59082-6_2

platforms on the territory of the Russian Federation in 2019[1]. Many scholars propose graph search approaches, but they leave out of account linguistic features of text collections [12, 15, 17]. The novelty of this study is that we propose a hybrid approach to detecting hidden communities which combines topic modeling and automatic topic labeling. Specific linguistic parameters of the posts are identified to provide the correct procedure of preprocessing texts. LDA is used for discovering topics distribution of users, topic labeling helps to improve topic interpretability.

VKontakte social network is the most attractive platform for contemporary researchers. Numerous algorithms were developed to detect close users based on graph searches [3, 19, 21]. However, the studies on automatic detection of hidden communities have never been carried out by complex methods.

2   Related Works

Most of the studies on detecting hidden communities on social networks use graph methods. In [12] hidden communities are detected in Facebook social network, the following algorithm being proposed.

1. At first, each node is considered a separate community. The objective function $L(M)$ is calculated:

$$L(M) = qH\left(Q\right) + \sum_{i=1}^{m} p_i H(P^i),\tag{1}$$

where

$$q = \sum_{i=1}^{m} q_i\tag{2}$$

is the probability of transition between communities at each step of a random walk, q_i is the probability of leaving community i,

$$p_i = \sum_{\alpha \in i} p_\alpha + q_i\tag{3}$$

shows the probability of staying in community i, p_α corresponds to the probability of getting into node α, Q is the number of communities, P^i is the number of nodes in community i. H corresponds to the entropy of the corresponding quantity.

2. Random walk forms a sequence of nodes.

3. Taking into account the frequency of occurrence of nodes, scholars select subsets of nodes in the resultant sequence, the subsets form communities.

4. For a given partition the metric $L(M)$ is calculated. If its value has become lower, then the partition M is preserved and the algorithm continues to work (returns to

---

[1] https://gs.statcounter.com/social-media-stats/all/russian-federation/2019.

step 2). In another case, if the value of the objective function $L(M)$ has not decreased, step 5 should be carried out.

5. The final partition M is considered the result of detecting a hidden community. In [5] there is another method: a function

$$H_\varepsilon(t) = \min\left\{ h : P[X(t) < h] \geq 1 - \varepsilon\left(\frac{\delta}{t^{\delta+1}}\right)\right\}, \tag{4}$$

is introduced, $X(t)$ shows the size of the largest constant component that can be selected during communication cycles (between 1 and t); h is the size of a hidden community that needs to be detected; ε is the confidence level, t shows the operating time of the cycles until $X(t)$ occurs. Upon reaching the level of values at which $X(t) \geq H(t)$, it can be claimed that a hidden community is detected at $1 - \varepsilon$.

In the paper [9] two stages of the detection of hidden communities in HICODE (Hidden Community Detection) algorithm are defined. The first stage consists in finding the number of community layers. The first layer is a group with the strongest and most obvious links, each subsequent layer having a lower degree of connections.

At the second stage called the refinement stage the quality of the layers has to be improved. This stage helps to obtain complete data because a stronger community structure can distort information about the structure of weaker communities.

The study [11] presents the results of the work on detecting hidden communities on the material of electronic versions of the Dialogue conference proceedings in 2000–2009. The results show topics that are popular among speakers and topics that will capture the attention of NLP researchers: artificial intelligence, speech synthesis, etc.

All the above described techniques emphasize the use of graph search, but they let linguistic properties of the texts out of sight.

3   Experiments with the Russian Corpus of VKontakte Posts

3.1   Corpus Collecting and Preprocessing

To test the approach, it was decided to select the posts of users with some parameters which are described below. The corpus of VKontakte Russian posts includes 8 679 402 tokens. The entire procedure of preprocessing consists of six stages: manual selection of users' accounts, creating a parser to scrape information about text posts, tokenization, deleting tokens using stop-list, adding bi- and trigrams, and manual post-editing.

The HSE developed VK Miner information system[2] for processing users' data, but the system does not meet the needs because it is unable to download text data of the posts. To create the parser, we used Python 3.7[3], BeautifulSoup4 library[4] for working with web data, as well as requests library. A user has only 20 posts on the wall, so it was decided to analyze the first ten pages and select text posts which were published no

---

[2] https://linis.hse.ru/soft-linis/.

[3] https://www.python.org/downloads/release/python-370/.

[4] https://www.crummy.com/software/BeautifulSoup/bs4/doc/.

earlier than 2018. We also took into consideration that texts, whose length is more than 200 characters, contain minimal semantic information, so further analysis could be carried out[5].

To implement further procedures, linguistic parameters of social network posts must be taken into account. In some cases users' texts are a combination of characters of different alphabets. For instance, most of the posts about films included characters from the Greek alphabet: from a graphic point of view, the Russian letter *M* is very similar to the Greek letter *M (mu)*. Users usually apply such replacement in order to "deceive the system" and avoid deleting a post due to copyright infringement. As a result of tokenization, the word *Машина (mashina)* was divided into two different tokens: *M* and *ашина (ashina)*. Such tokens were obtained from most words represented by a number-letter complex or a complex of letters of various alphabets, therefore, these words were post-edited manually or added to a stop-list.

The most common difficulties with tokenization of post texts are the following:

- a combination of letters of different alphabets;
- the use of hieroglyphics and symbols of the Arabic alphabet;
- the use of diacritics, which were also removed from the texts of posts;
- the use of non-text characters (icons, etc.).

To add a token to the processed corpus, it has to be lemmatized using PyMorphy2 library[6]. The module extracts the normal form of a token from OpenCorpora resource[7] and automatically assigns it to the token. Lemmatization of both dictionary and non-dictionary forms was carried out. During processing non-dictionary forms, a predictor was used for words related to the names of social networks, neologisms, occasionalisms, unknown surnames and proper names, as well as morphologically and graphically adapted foreign words. If a word was not found in the dictionary, the module analyzes it by analogy with many other similar tokens. In most cases the correct normal forms were assigned, but some forms turned out to be erroneous and required further post-editing. The most frequent erroneous normal forms are the following:

- *инстагра (instagra)* lemma was assigned to Instagram social network, the module recognized the derived word as a feminine dative case form of a plural noun (compare *мамам – мама (mamam – mama), играм – игра (igram – igra)*, etc.);
- the word *тревел (travel)* was recognized as a past tense form of the verb *треветь (trevet')*;
- the noun *море (more)* in the nominative case was recognized as a form of the prepositional singular form of the word *мор (mor)*;
- the noun *фейк (feik)* has taken on the form *фейка (feika)* with the meaning *little fairy*.

---

[5] https://blog.hubspot.com/marketing/character-count-guide.

[6] https://pymorphy2.readthedocs.io/en/latest/.

[7] http://opencorpora.org/.

During lemmatization, each lexeme was checked for its presence in the stop-list which includes prepositions, conjunctions, particles, interjections, symbols of various alphabets, obscene vocabulary, abbreviations, etc. The total number of stop words is 1423.

Finally, we used gensim[8] library to add bi- and trigrams to our corpus that can help to make topic models more informative: some of them are *рандомный_игра (random_game)*, *добрый_дело (good_deed)*, *театр_нидерланды (theatre_Netherlands)* and others.

After preprocessing the corpus size turned out to be 2 352 979 tokens (714 users, 25 768 posts).

3.2 Author-Topic Models

After collecting the corpus of Russian social networks posts, it is necessary to create and train an author-topic model (ATM). The author-topic model combines a topic model that represents the links between words, documents and topics, and an author model that represents the links between authors and documents. Figure 1 shows a scheme of ATM.

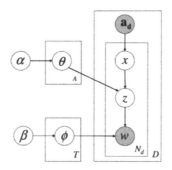

Fig. 1. Visual representation of ATM.

In the above scheme, α and β are the Latent Dirichlet Allocation (LDA) parameters, θ is the topic distribution of documents, z corresponds to the topic of each word in the document, χ is the author of each word in the document, ϕ is the distribution of words in each topic.

ATM is used to solve a number of tasks: for instance, to identify the gender or age of authors of posts on social networks, to determine the authorship of anonymous texts, etc. Papers [4, 20, 23] contain a more detailed description of ATM applications.

In our study, ATM was built by means of gensim library adapted to the conditions of our experiment. The choice of this library is explained by the compatibility with

---

[8] https://radimrehurek.com/gensim/.

third-party libraries for data visualization. The whole procedure of ATM training consists of three stages.

1. Search for the most optimal number of topics for each user's subcorpus using the U-Mass measure as it reflects topic coherence value which is treated as a level of human interpretability of the model based on relatedness of words and documents within a topic:

$$core(v_i, v_j, \epsilon) = \log\frac{D(v_i, v_j) + \epsilon}{D(v_i)}, \tag{5}$$

where $D(v_i, v_j)$ shows the number of documents that contains v_i and v_j words, $D(v_i)$ is the number of documents containing v_i words [22]. The highest value of $core(v_i, v_j, \epsilon)$ shows that ATM will include the appropriate number of topics. A series of experiments was conducted to identify the optimal number of topics, the following parameters being used: the minimum number of topics was 5, the maximum was 35, and the step was 5. As a result, graphic data were obtained for further ATM training. In Fig. 2 there is an example of such graphic data.

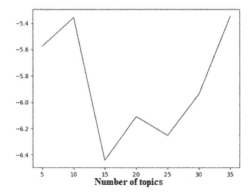

Fig. 2. Optimal number of posts for user 385450.

2. Training ATM with the obtained number of topics for users, each topic containing 10 lemmata with corresponding weights. PyLDAvis library[9] was also used for visualizing topic models.

Topics of user 385450

Topic 1: *тип, проблема, психологический, противоположность, интровертный, квантовый, установка, экстравертный, год, общий (type, problem, psychological, opposite, introversive, quantum, extroversive, year, common).*

Topic 2: *письмо, книга, письменность, библиотека, электронный, древний, год, мир, сайт, стать (letter, book, writing, library, electronic, ancient, year, world, site, become).*

---

[9]   https://github.com/bmabey/pyLDAvis.

Topic 3: *эрмитаж, музей, коллекция, отдел, искусство, дворец, здание, год, центр, главный (hermitage, museum, collection, department, art, palace, building, year, center, main).*

Topic 4: *цивилизация, год, эйнштейн, квантовый, картина, мир, человек, место, город, религия (civilization, year, einstein, quantum, picture, world, person, place, city, religion).*

Topic 5: *искусство, век, год, древний, страна, книга, италия, история, испания, англия (art, century, year, ancient, country, book, italy, history, spain, england).*

Topic 6: *год, письмо, письменность, азия, тип, мозг, книга, военный, человек, жизнь (year, writing, writing, asia, type, brain, book, military, person, life).*

Topic 7: *страна, эпоха, развитие, модерн, жизнь, производство, будущее, мир, считаться, модель (country, epoch, development, modern, life, production, future, world, reckon, model).*

Topic 8: *год, красота, азия, военный, война, южный, восточный, индия, рим, империя (year, beauty, asia, military, war, south, east, india, rome, empire).*

Topic 9: *мозг, глава, сражение, битва, книга, человек, человеческий, наука, разум, космос (brain, chapter, fighting, battle, book, man, human, science, mind, space).*

Topic 10: *мозг, человек, компьютер, язык, информация, знать, время, мочь, делать, искусство (brain, human, computer, language, information, know, time, can, do, art).*

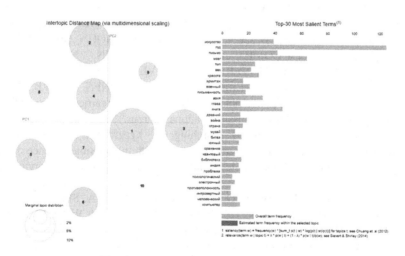

Fig. 3. Topic distribution of user 385450.

In Fig. 3 there is a topic distribution of user 385450. Topic distribution is connected with the selection of a specific metric for visualization. T-distributed Stochastic Neighbour Embedding (tsne) method as a metric for multidimensional scaling was chosen to compress data and represent multidimensional variables.

3. Saving the results in a *.txt file, which will be used for the further procedures.

3.3   Automatic Labeling of Topics

While working with algorithms for topic modeling, there is often a need to improve the interpretability of the obtained topics, so automatic topic labeling procedures are used. Most of the papers describing various approaches to the implementation of these algorithms were carried out on the material of English text corpora [1, 2, 15, 16], the Russian material is not fully investigated. Studies on Russian data [8, 14] discuss the procedure of searching for topic labels using the morphosyntactic parsing algorithm in Yandex searching engine and using the Explicit Semantic Analysis (ESA) algorithm applied to the Wikipedia vector space model. Experiments with the corpus of linguistic texts have been successfully carried out. However, while working with the corpus of social networks, it should be noted that using Wikipedia data do not seem appropriate to us, since this resource is more focused on scientific texts, not on texts of posts on social networks. It should also be noted that pages of users can contain texts of different styles: someone writes about politics, someone writes about his personal life.

The topic labels are expected to be hypernyms, holonyms, or abstract lexical units that generalize and explain the meaning of the selected topic words [23]. In our study, the procedure of topic labeling includes several stages.

1. Labels from external sources. We created a parser which automatically extracted topic words and created a search query for Google. The choice of this web search engine is explained by the fact that the query results are ranked using PageRank. A great number of repeated requests can lead to ban, so User-Agent module was used for specifying versions of operating systems and browsers. We extracted collocations of site headers using Pymorphy2 and the following patterns: ADJF + NOUN, NOUN + NOUN_GEN, NOUN + NOUN_GEN + NOUN_GEN, PREP + NOUN + NOUN_GEN, NOUN + PREP + NOUN, NOUN + COMP, ADVB + ADJF + NOUN, NOUN + CONJ + NOUN, ADJF + ADJF + NOUN, VERB + ADVB, VERB + INFN, PRTF + NOUN, PRTS + NOUN, INFN + ADVB, VERB + PREP + NOUN, INFN + ADVB, PREP + ADJF + NOUN. The ranking of the obtained collocations was carried out according to the frequency of occurrence in the Russian National Corpus[10].

2. Obtaining label candidates by searching for hypernyms in RuWordNet[11]. Restrictions on the use of RuWordNet are connected with the possible absence of topic words in the thesaurus and the limited number of their potential hypernyms. The hypernyms were ranked in accordance with ipm frequencies from a Frequency Dictionary of Contemporary Russian by O.N. Lyashevskaya and S.A. Sharov[12].

3. Labels from internal sources. To search for labels, we built Word2Vec CBOW model of the corpus with the help of gensim library. The context window was 5, the minimum word frequency for including in the model was also 5, dimensionality of the word vectors was 100. For all ten topic words we created an average topic vector, the five closest associates being selected according to the vector representation of words. These associates can also be considered as candidates for topic labels.

---

[10]   http://www.ruscorpora.ru/new/.

[11]   http://www.ruwordnet.ru/ru.

[12]   http://dict.ruslang.ru/freq.php.

4. The last stage was comparison of the obtained search results, manual post-processing and assignment of labels to topics. As a result, labels have improved the interpretability of topic models and models of hidden communities. Below we present a comparative table of candidates for topic labels (Table 1).

It should be noted that the first word or collocation may turn out to be the final label. However, according to the results of extracting external and internal labels, the first word does not describe a topic. The combination of applied methods for extracting labels improves the interpretability of users' topics, in the future, a quantitative analysis of these procedures will be required.

Table 1. Examples of the results of automatic topic labeling.

User	Topic	Labels from external sources (Google and RuWordNet)	Labels from internal sources (Word2Vec)
834	адрес, внимание, **страница,** надёжно, ссылка, взламывать, скоро, перевод, **фотография**, страничка (address, attention, **page**, reliable, link, hack, soon, translation, **photo**, page)	Россия Москва, информационная безопасность, **рядовые пользователи**, победитель, **изображение, фрагмент** (Russia Moscow, **information security**, ordinary users, conqueror, **image, fragment**)	**ссылка, видео**, просить, **фотография**, появиться (**link**, **video**, ask, **photo**, appear)
57296	конкурс, **университет,** государственный, молодёжный, проект научный, российский, год, имя, студенческий (competition, **university**, state, youth, scientific project, Russian, year, name, student)	новости науки, **Российская ассоциация, учебное заведение**, соревнование, опыт, молодежный (science news, **Russian Association, educational institution**, competition, youth experience)	октябрь, июнь, белгород, миллиард, **институт** (october, june, belgorod, billion, **institute**)
5019159	**съёмка**, ребёнок, заранее, родитель, фото, воспитатель, одежда, **снимок, коллаж**, группа (**shooting**, child, in advance, parent, photo, educator, clothes, **snapshot, collage**, group)	в детском саду, личный блог, детская **фотосъемка, изображение**, снять, состав (in kindergarten, personal blog, **children photography image**, take a picture, structure)	прекрасный, **образ**, радость, здоровье, сад (beautiful, **image**, joy, health, garden)

3.4   Model of Hidden Communities in VKontakte Social Network

According to the results of the algorithm, data on the topic organization of the corpus were structured, we took into account labels received from external and internal sources. A table of links between topics and user posts was compiled. Before building the model, it was necessary to combine the lexical-semantic variants of labels – the names of topics.

1. When finding adjective and noun modifiers in collocations, we chose the first one as a dominant collocation: *student organizations, organizations of students – student organizations*.

2. Antonymic topics were combined into one, since they are based on a single semantic component: *diseases, health – healthy life*.

3. Synonyms are words in the same semantic field, so they were combined into one label: *films and series, cinema, television, film production – film production; religion, Orthodox councils – religion; photography, professional photo shoots – photographing*.

Gephi application[13] was used to create models of hidden communities. In Fig. 4 we present a comparison of the models built on the same dataset: a) communities detected with the help of the hybrid method, b) communities detected with the help of a graph-based method and Jaccard similarity index. The hybrid model allows to reveal a fine-grained structure of the dataset while the graph-based model gives a fragmentary description of relations existing in the corpus.

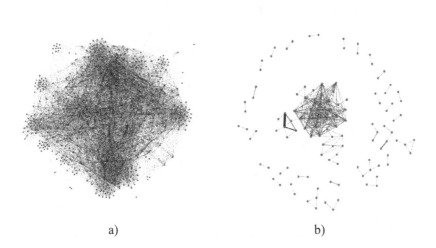

a)   b)

Fig. 4. Models of hidden communities.

---

[13]   https://gephi.org/.

Hidden communities in the given models are represented as a group of nodes connected by edges of the same colour. We focus on the hybrid model which unites 714 users forming 60 hidden communities, they cover a large number of topics, from travelling and health to narrowly focused topics, such as the Ukraine and the Crimea.

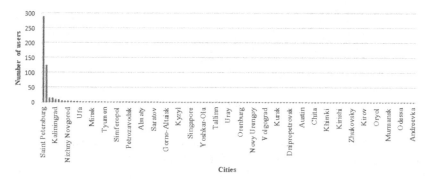

Fig. 5. Users' distribution by places of residence.

Figures 5, 6 and 7 show a quantitative analysis of users. After downloading additional users' data with the help of vk_api library[14] (name, age, place of residence, groups and subscriptions), it was found that many of the users live in large cities, but there are users who live in different parts of Russia: for instance, a user with ID 820468 lives in Tula, and a user with ID 70978256 lives in St. Petersburg, they being members of a hidden community dedicated to festivals. The average age of the users is 40 years, at the same time 365 users hide the years of birth.

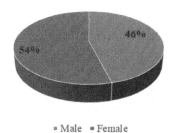

▪ Male ▪ Female

Fig. 6. Users' distribution by gender.

---

14   https://vk-api.readthedocs.io/en/latest/.

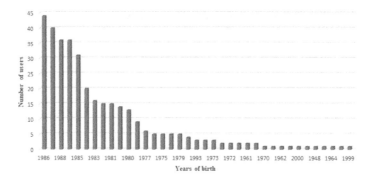

Fig. 7. Users' distribution by years of birth.

Graph methods allow to determine the distance between members of hidden communities. One can check the theory of the rule of six handshakes put forward by American psychologists Milgram and Travers. The idea of the theory is that all the people are six, or fewer, social connections away from each other [7, 10]. In 2018 in honor of its birthday VKontakte gave access to the application for several days which could help to check the theory[15]. In this study we chose the online service[16] which allows to visualize the distance between users on the network. In Fig. 8 users 3854113 and 4817853 from "literature" hidden community are at a distance of two friends from each other (the centers of the graphs are the objective users). If the two user graphs have no links, it means that their friends are not explicitly connected with each other and it is necessary to add a third user to determine the distance between the users.

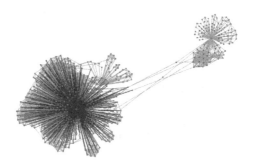

Fig. 8. Distance between users 3854113 and 4817853.

When analyzing overlapping groups of hidden communities members, we observed that the most frequent user groups are humorous and educational (Fig. 9): *Безумный юмор (Crazy sence of humour), СССР – Вспомним всё (The USSR – Let's recall everything), etc.*

```
Безумный юмор {'103825', '160319519'}
Приколы {'32133', '40585', '1825558', '75944', '103825'}
СССР - Вспомним всё {'103825', '160319519'}
И это факт! (PURE FACTS) {'103825', '325072'}
Радио Юмор FM {'298245303', '103825', '2500528', '2163686'}
МИР РУКОДЕЛИЯ {'387174', '1218605', '161', '103825', '4644495', '1395179'}
Экология России {'1208871', '249543643', '2003892', '103825', '4644495', '965362', '568003'}
Радио Romantika {'103825', '232582', '2796357', '2163686'}
```

Fig. 9. Examples of overlapping communities.

Members of hidden communities are rarely members of the same real topic communities. User 878 is a member of such communities as *ВКонтакте для сообществ и SMMщики (VKontakte for Communities and SMM community members)*, he is a member of "digital environment" hidden community. User 190868, who is also a member of this hidden community, is a member of other communities: for instance, *ЮТУБЕР (YouTuber)*. Both of them are members of *Стартапы и бизнес (Startups and business)* community, however, our algorithm did not include them in "business" hidden community. It can be explained by the fact that members of real communities are unlikely to make reposts from the communities in which they are less interested.

4 Results and Evaluation

Figures 10 and 11 present the distribution of hidden communities by users and the distribution of users by hidden communities. The largest groups were communities on health and education, while communities on China, comics and self-development are less popular. Almost half of all the users are in one community, only 3% of users are not in hidden communities. Less than 1% of users are the members of five hidden communities.

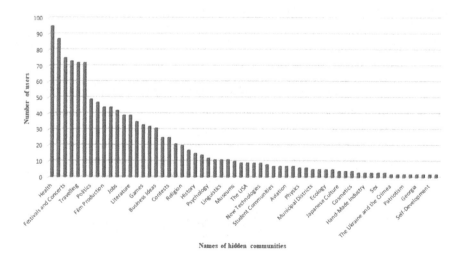

Fig. 10. Distribution of hidden communities by users.

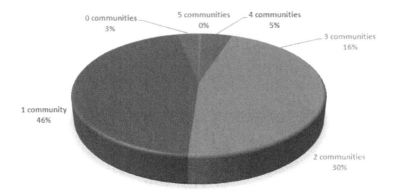

Fig. 11. Distribution of users by hidden communities.

While assessing the quality of procedures, it is worth noting its advantages. Topic modeling algorithms have already been successfully adapted both to Russian corpora of posts on social networks and Russian corpora of scientific literature [6, 13, 18, 24, 25]. When working with labels, we used a double ranking algorithm (PageRank in Google and ipm frequencies in the Russian National Corpus and a Frequency Dictionary of Contemporary Russian by O.N. Lyashevskaya and S.A. Sharov) for avoiding low-frequency labels.

We faced specific difficulties in course of posts preprocessing. This problem is connected with the graphical and spelling ways of writing a post, which lead to constant improving the procedures of tokenization and normalization.

As for automatic topic labeling, there is no gold standard for assessing the quality of labeling procedures, it was decided to create a survey using Google Forms service, the results helped to assess automatically generated candidates for topic labels. We asked 10 independent assessors to analyze the high-frequency n-gram labels described in 3.3. They were supposed to set 1 to relevant candidates which can be used as topic labels and 0 if candidates were irrelevant. Below we present some results of assessing, cf. Tables 2 and 3.

Table 2. Results of an expert assessment of candidates for topic labels of user 41003889.

Topic of user 41003889: *овсянка, банка, ложка, йогурт, молоко, крышка, стакан, закрывать, добавлять, мёд (oatmeal, jar, spoon, yogurt, milk, cap, glass, close, add, honey).*		
Bi- and trigrams	*рецепты приготовления (cooking recipes)*	*запаренная овсянка (steamed oatmeal)*	*великолепные завтраки (amazing breakfasts)*
Assessing results	10	6	9
Unigrams	*крупа (groats)*	*кулинария (cookery)*	*искусство (art)*
Assessing results	3	6	0

Table 3. Results of an expert assessment of candidates for topic labels of user 820468.

Topic of user 820468: *человек, джокер, оказаться, обстоятельство, убить, язык, несколько, изучение, выбрать, жертва (man, joker, turn out, circumstance, kill, language, several, study, choose, victim).*			
Bi- and trigrams	*лучший фильм (the best film)*	*фильм года (film of the year)*	*российские фильмы (Russian films)*
Assessing results	9	7	0
Unigrams	*существо (creature)*	*событие (event)*	*отклонение (deviation)*
Assessing results	2	4	4

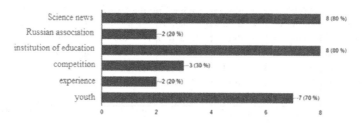

Fig. 12. Comparative diagram of candidates for topic labels (user 57296, topic words: *competition, university, state, youth, project, scientific, Russian, year, name, student*).

In Fig. 12 we present the results of the assessment. In the example, the assessors almost unanimously agree that the *science news* and *institute of education* are the best labels of the topic, while *experience* and *competition* labels have poor results.

Despite the fact that unigram candidates for labels were assessed lower than bi- and trigrams, it is worth noting that the unigrams are suitable for more general topics, meanwhile bi- and trigrams should be used for more narrowly focused topics. The combination of expert-based and automatic procedures helps to improve the interpretability of topic models.

5 Summary

Many researchers, when building models for detecting hidden communities, often implement graph search. Combining topic modeling and automatic topic labeling allows to fill in the elaboration of hidden community detection.

We developed a hybrid algorithm for detecting communities in Russian segment of social networks, described the advantages and disadvantages and compared it with existing methods. The results make it possible to obtain new information on possible social groups on the Internet. Evaluation of the procedures showed that, despite minor difficulties, it works as well as methods based on mathematical models.

Using data from social networks allows to solve a number of related tasks: for instance, tracking trends among users and, as a result, creating relevant content. Further studies will be connected with the increasing the corpus including different Russian online platforms, so it will be possible to detect hidden communities of users in various social networks.

References

1. Aletras, N., Stevenson, M.: Representing topics using images. In: NAACL-HLT 2013, pp. 158–167 (2013). https://www.researchgate.net/publication/236142659_Representing_Topics_Using_Images

2. Allahyari, M., Pouriyeh, S., Kochut, K., Arabnia, H.: A knowledge-based topic modeling approach for automatic topic labeling. Int. J. Adv. Comput. Sci. Appl. **8**, 335–349 (2017). https://doi.org/10.14569/IJACSA.2017.080947

3. Alymov, A., Babiychuk, G.: Analysis of user profiles to determine the most significant objects of a subset in VK social network. Sci. Technol. Educ. **10**, 41–45 (2017)

4. Argamon, Sh., Koppel, M., Pennebaker, J., Schler, A.: Automatically profiling the author of an anonymous text. Commun. ACM – Inspiring Women Comput. **52**, 119–123 (2009). https://doi.org/10.1145/1461928.1461959

5. Baumes, J., Goldberg, M., Magdon-Ismail, M., Wallace, W.A.: Discovering hidden groups in communication networks. In: Chen, H., Moore, R., Zeng, D.D., Leavitt, J. (eds.) ISI 2004. LNCS, vol. 3073, pp. 378–389. Springer, Heidelberg (2004). https://doi.org/10.1007/978-3-540-25952-7_28

6. Bodrunova, S., Blekanov, I., Kukarkin, M.: Topics in the Russian Twitter and relations between their interpretability and sentiment. In: Sixth International Conference on Social Networks Analysis, Management and Security, pp. 549–554 (2019). https://doi.org/10.1109/snams.2019.8931725

7. Buchanan, M.: Nexus: Small Worlds and the Groundbreaking Science of Networks, 235 p. W.W. Norton & Company (2002)

8. Erofeeva, A., Mitrofanova, O.: Automatic topic labeling in topic models of the corpora of Russian texts. Struct. Appl. Linguist. **12**, 122–147 (2016)

9. He, K., Li, Y., Soundarajan, S., Hopcroft, J.: Hidden community detection in social networks. Inf. Sci. **425**, 92–106 (2018). https://doi.org/10.1016/j.ins.2017.10.019

10. Ipatov, Yu., Kalagin, I., Krevetsky, A., Sokolov, B.: Analysis of dynamic characteristics of complex graph structures. In: News of higher educational institutions. Instrum. Eng. **6**, 511–516 (2018). https://doi.org/10.17586/0021-3454-2019-62-6-511-516

11. Khoroshevsky, V., Efimenko, I.: Semantic technologies in scientometrics: tasks, problems, solutions and prospects. In: Cognitive-Semiotic Aspects of Modeling in Humanities, pp. 221–267 (2017)

12. Kolomeychenko, M., Chepovsky, A., Chepovsky, A.: Algorithm for detecting communities in social networks. Fundam. Appl. Math. **19**, 21–32 (2014)

13. Koltsov, S., Pashakhin, S., Dokuka, S.: A full-cycle methodology for news topic modeling and user feedback research. In: Staab, S., Koltsova, O., Ignatov, D.I. (eds.) SocInfo 2018. LNCS, vol. 11185, pp. 308–321. Springer, Cham (2018). https://doi.org/10.1007/978-3-030-01129-1_19

14. Kriukova, A., Erofeeva, A., Mitrofanova, O., Sukharev, K.: Explicit semantic analysis as a means for topic labelling. In: Ustalov, D., Filchenkov, A., Pivovarova, L., Žižka, J. (eds.) AINL 2018. CCIS, vol. 930, pp. 110–116. Springer, Cham (2018). https://doi.org/10.1007/978-3-030-01204-5_11

15. Magatti, D., Calegari, S., Ciucci, D., Stella, F.: Automatic labeling of topics. In: Intelligent Systems Design and Applications, pp. 1227–1232 (2009). https://doi.org/10.1109/isda.2009. 165. https://www.researchgate.net/publication/216827237_Automatic_Labeling_Of_Topics

16. Mei, Q., Shen, X., Zhai, C.: Automatic labeling of multinomial topic models. In: SIGKDD, pp. 490–499 (2007). https://doi.org/10.1145/1281192.1281246

17. Mityagin, S., Yakushev, A., Bukhanovsky, A.: Research on Internet social networks for detecting the related interests of drug addicted people. Engineering **6**, 59–64 (2012)

18. Nikolenko, S., Koltcov, S., Koltsova, O.: Topic modelling for qualitative studies. J. Inf. Sci. **43**, 88–102 (2015). https://doi.org/10.1177/0165551515617393

19. Nitkin, D., Yudina, M.: Programme for extracting information on relationships of users of VK social network. Dyn. Syst. Mech. Mach. **4**, 32–35 (2016)

20. Panicheva, P., Mirzagitova, A., Ledovaya, Y.: Semantic feature aggregation for gender identification in Russian Facebook. In: Filchenkov, A., Pivovarova, L., Žižka, J. (eds.) AINL 2017. CCIS, vol. 789, pp. 3–15. Springer, Cham (2018). https://doi.org/10.1007/978-3-319-71746-3_1

21. Smirnova, O.: Risk estimation for vk.com accounts exposed to suicide-themed quests. In: Modern Education Technologies and IT-Education, vol. 13, pp. 53–60 (2017)

22. Stevens, K., Kegelmeyer, P., Andrzejewski D., Buttler D.: Exploring topic coherence over many models and many topics. In: Proceedings of the 2012 Joint Conference on Empirical Methods in Natural Language Processing and Computational Natural Language Learning, pp. 952–961 (2012)

23. Mitrofanova, O., Sampetova, V., Mamaev, I., Moskvina, A., Sukharev, K.: Topic modelling of the russian corpus of pikabu posts: author-topic distribution and topic labelling. In: Proceedings of the International Conference « Internet and Modern Society» (IMS 2020), International Workshop «Computational Linguistics» (CompLing-2020) (2020, in press)

24. Vorontsov, K., Frei, O., Apishev, M., Romov, P., Suvorova, M., Yanina, A.: Non-Bayesian additive regularization for multimodal topic modeling of large collections. In: Proceedings of the 2015 Workshop on Topic Models: Post-Processing and Applications, pp. 29–37 (2015). https://doi.org/10.1145/2809936.2809943

25. Vorontsov, K., Potapenko, A.: Tutorial on probabilistic topic modeling: additive regularization for stochastic matrix factorization. In: Ignatov, D.I., Khachay, M.Y., Panchenko, A., Konstantinova, N., Yavorskiy, R.E. (eds.) AIST 2014. CCIS, vol. 436, pp. 29–46. Springer, Cham (2014). https://doi.org/10.1007/978-3-319-12580-0_3

Dialog Modelling Experiments
with Finnish One-to-One Chat Data

Janne Kauttonen(✉) [iD] and Lili Aunimo [iD]

Haaga-Helia University of Applied Sciences, Ratapihantie 13, 00520 Helsinki, Finland
{janne.kauttonen,lili.aunimo}@haaga-helia.fi

Abstract. We analyzed two conversational corpora in Finnish: A public library question-answering (QA) data and a private medical chat
dataWe developed response retrieval (ranking) models using TF-IDF,
StarSpace, ESIM and BERT methods. These four represent techniques
ranging from the simple and classical ones to recent pretrained transformer neural networks. We evaluated the effect of different preprocessing strategies, including raw, casing, lemmatization and spell-checking for
the different methods. Using our medical chat data, We also developed
a novel three-stage preprocessing pipeline with speaker role classification. We found the BERT model pretrained with Finnish (FinBERT) an
unambiguous winner in ranking accuracy, reaching 92.2% for the medical
chat and 98.7% for the library QA in the 1-out-of-10 response ranking
task where the chance level was 10%. The best accuracies were reached
using uncased text with spell-checking (BERT models) or lemmatization (non-BERT models). The role of preprocessing had less impact for
BERT models compared to the classical and other neural network models. Furthermore, we found the TF-IDF method still a strong baseline for
the vocabulary-rich library QA task, even surpassing the more advanced
StarSpace method. Our results highlight the complex interplay between
preprocessing strategies and model type when choosing the optimal approach in chat-data modelling. Our study is the first work on dialogue
modelling using neural networks for the Finnish language. It is also first
of the kind to use real medical chat data. Our work contributes towards
the development of automated chatbots in the professional domain.

Keywords: NLP · Deep learning · Chatbots · Information retrieval ·
Multi-turn conversation · Dialogue modelling

1 Introduction

Lately there has been an increase in the amount of computer-mediated written communication [4]. Most of this communication happens between humans,
but an increasingly large amount of it is taking place between a human and a
bot. Examples of professional synchronous computer-mediated communication
between humans are different chat services offered by companies' customer care,
healthcare or public organizations. On the other hand, people are also using chat

© Springer Nature Switzerland AG 2020
A. Filchenkov et al. (Eds.): AINL 2020, CCIS 1292, pp. 34–53, 2020.
https://doi.org/10.1007/978-3-030-59082-6_3

services in their private life to communicate with friends and different communities. This study concentrates on professional one-to-one chat services offered to private people by companies or public organizations. The aim is to model the dialogues with such an accuracy that allows at least partial automation. The automation only concerns the utterances of the professionals and not those of the customer. In this work we use the terms *chat*, *conversation* and *dialogue* interchangeably. Many researchers use the term chat to denote chit-chat type of general domain conversation without a specific goal [4]. However, chat is the term often used when companies or other organizations offer customer care or other services through a synchronous written digital channel.

The need for dialog modelling of one-to-one chat data becomes more and more urgent as the amount of professional computer-mediated communication increases because the need for automating the routine parts of the conversation grows. This not only makes the work of the professional more interesting but also increases the productivity of his/her work. Dialog modelling and thus also automation of expert dialogues becomes more feasible because as the amount of available chat data grows. Even though the amount of chat data is vast, it is in general not publicly available and the work done on modelling one-to-one chat dialogues for minor languages is scant. Most work concentrates on the English and Chinese languages. However, also rarer languages, such as the Finnish language, would benefit from dialogue models that could be used to partially automate corporate chat services.

This study attempts to fill the research gap on dialog modelling for Finnish chat data. It uses two datasets to build and compare four different retrieval-based models. Also different preprocessing schemes are studied. The first dataset is the "Ask a librarian" single-turn one-to-one chat corpus and the second is the multiturn one-to-one medical chat corpus. Both datasets consist of synchronous goal-oriented dialogues. *Multi-turn dialogue* consists of more than one turn per speaker, while *single-turn dialogue* contains only one (here a question-answer pair). One turn may consist of one or more *utterances*. *One-to-one chat* means that the dialogue only has two conversationalists as opposed to a *multicast chat* that has more than two participants. A *goal-oriented dialogue* is guided by the goal that one of the conversationalists attempts to fulfill [24]. A chit-chat dialogue is an example of a chat without a goal.

Both retrieval-based and generative chatbots have been implemented using a data set of existing live chat dialogues between humans [33]. Some chatbots only take into account the last utterance and search for an answer based on it [30]. This type of chatbots are called single-turn chatbots or question answering systems. The Library QA data modelled in this study represents this type of dialogue system. Other chatbots, like the medical chat presented in this paper, take as input the whole history (aka context) of the chat. These are called multi-turn chatbots [33]. The following sections of this paper contain a review of the related work, the description of the data and methods used in the research, presentation of the results and finally an analysis of their significance.

2   Related Work

2.1   Language Resources for One-to-One Chat Dialogue Data

Data sources for analyzing written Finnish one-to-one dialogue data, also called synchronous computer-mediated communication (CMC) or simply chat data, are scarce. There is only one publicly available data set and it contains only single-turn conversation data. This is the "Ask a Librarian data set" that has been used in the experimental part of this study. This data set has been available already since 1999 [21]. The questions and answers can be freely scraped from the web page of the service and contains over 40,000 question-answer (QA) pairs. There is almost no previous research done on this data set. Only some thesis work on the classification of question and answer types of the data have been done from the point of view of information retrieval systems [21].

The Language Bank of Finland[1] has some dialogue-related corpora in Finnish. These include the Suomi24 Sentences Corpus [1], which contains all the discussion forums of the Suomi24 online social networking website from a time period of 17 years. However, the data from a discussion forum is very different when compared to the one-to-one dialog data that the present study addresses. In the study at hand, the one-to-one discussions happen between a professional (expert) in the field (a medical doctor or a librarian) and a customer (typically a layman). A similar setting can be found in several professional use cases of one-to-one chat such as: a pedagogical chat involving a teacher and a student, customer care chat involving a company's representative and a customer or public organization's chat service involving a public servant and a citizen. According to our knowledge there are no public data sets in the Finnish language for these use cases.

For some globally significant languages, such as English and Chinese, there are publicly available corpora containing multi-turn one-to-one dialogues. For English, one of the best-known dialogue corpora is the Ubuntu Dialogue Corpus [16] with 930,000 dialogues scraped from the Ubuntu operating system support IRC channel. The corpus consists of one-to-one dialogues where one dialog participant is asking for support on an Ubuntu-related problem and the other participant is answering to this information need. For the Chinese language, there is the Douban Conversation Corpus that contains 1.1 million one-to-one dialogues longer than 2 turns from the Douban Group, which is a popular social networking site in China [33]. In contrast to the domain specific Ubuntu Dialogue Corpus, this data contains open domain conversations. The ParlAI python framework [20] also contains publicly available English language dialogue data sets. The framework includes over 80 different data sets that can be used for dialogue modelling. They range from open-domain chit-chat to visual question answering.

Some European languages also have publicly available chat corpora. For German, there are two well-known chat corpora: the Düsseldorf CMC Corpus and

---

[1]   https://www.kielipankki.fi.

the Dortmund Chat Corpus [4]. They include annotated chat dialogues from several professional contexts such as teaching, learning and counselling. For the French language, there is the French version of the Ubuntu Corpus [22]. As publicly available one-to-one multi-turn chat corpora are rare in many languages, researchers have used the Web as a corpus. For example, it is technically possible to scrape dialogue data from any discussion forum, Twitter or for example Wikipedia Talk pages [3]. Twitter data has been used as a data source to model dialogue [23]. However, the authors report that most of the dialog data on Twitter consisted of only two-turn dialogues.

2.2   Machine Learning for Chat Data Modelling

The existing approaches to dialogue modelling include retrieval-based methods and generation-based methods [25, 26, 32]. Retrieval based methods select a proper response for the current conversation from a repository with response selection algorithms, and have the advantage of producing informative and fluent responses [10]. As the responses are predefined, the pool of responses is limited, but one has full control what the model can return. On the other hand, generation-based models maximize the probability of generating a response given the previous dialogue. This approach enables the incorporation of rich context when mapping between consecutive dialogue turns [10]. The responses are not predefined and can be better matched to the context/question, however there is less control over the responses.

The experimental part of this study is based on retrieval-based methods. The performance of four different methods is tested against two different Finnish one-to-one chat dialogue datasets. The methods range from the basic TF-IDF method to massive pretrained transformer networks (BERT). The methods are described in the following.

TF-IDF Similarity. This classic method is based on computing occurrences of words within and between samples. The method is widely used in information retrieval [17]. It is often used as a baseline method in retrieval-based dialogue models [16]. A TF-IDF-based model was also used in previous work in Finnish language question answering [2]. The TF-IDF model is typically used to model how important a term is in a given document. When modelling dialogue data, the model is used to capture the importance of a term in a given context [16]. The term frequency (TF) is simply the number of times the term appears in the context. The inverse document frequency (IDF) is the total number dialogues divided by the number of dialogues the term appears in. To retrieve the next turn in a dialogue, the TF-IDF vectors are first calculated for the context and for each response candidate in the set of candidate responses. The response with the highest cosine similarity with the context is then selected as the output.

Word2vec, fastText and StarSpace. Word2vec [19], fastText [14] and StarSpace [31] represent different generations of single hidden-layer (aka shallow) neural network models based on *embeddings*, i.e., dense vector representations of features (e.g., words and their n-grams). The latter two methods can be used both unsupervised, trained on unlabeled text, and supervised manner from series of labelled documents. StarSpace, the most recent of the three, also allows ranking of documents based on their similarity. fastText and StarSpace are lightweight and fast to train and can be considered as powerful baselines for various supervised language tasks [13,31].

ESIM. Enhanced Sequential Inference Model (ESIM) is a deep neural network model introduced by Chen et al. [7] and based on embeddings, bidirectional Long-Short Term Memory (bi-LSTM) cells and attention mechanism. The base model was further expanded by Dong et al. [9] to include character level embeddings to allow better handling of out-of-vocabulary words. The network contains two inputs (context and response) and consists of word representation (both token and character-wise), context representation (bi-LSTM), attention matching, matching aggregation (bi-LSTM), mean pooling, input concatenation and prediction layers. The sigmoid output results in 0/1 for perfectly mismatched/matched context and response. The ESIM model has been reported being among the best for the Ubuntu retrieval task [6,13].

BERT. The Bidirectional Encoder Representations for Transformers (BERT) model by Devlin et al. [8] is the current state-of-the-art language model for a wide range of natural language processing tasks, such as named entity recognition and classification in English [8]. BERT is a deep neural network based on bidirectional transformer encoders and multi-head attention mechanism described in [27]. The models used in this work contained 12 layers with 12 attention heads.

For Finnish, the original model is the multilingual BERT[2] (ML BERT) trained on pooled data from 102 (uncased) or 104 (cased) languages, one being Finnish. Recently, Virtanen et al. [29] released a FinBERT model trained on solely for Finnish. This new model was found superior to ML BERT in parsing and classification tasks for Finnish.

2.3   Evaluation Results for Chat Data Models

The results of chat data models are typically evaluated by formulating the task as best response selection given the context. This approach requires no human labels as the test data is formed by taking a set of data aside for testing purposes and by creating using random sampling a dataset where one response is correct and the rest are incorrect [16]. A common evaluation metric is to report the accuracy for choosing the correct response. With 1 correct and 9 incorrect responses, this evaluation measure is called *1 in 10 R@1* where 1 in 10 means

---

[2] https://github.com/google-research/bert.

that one must choose 1 correct response out of 10 candidates and *R@1* simply denotes *recall@1* [16,26]. The chance level (random selection) for 1 in 10 R@1 score is 0.10. *Recall@k* is a commonly used metric in information retrieval tasks. It means that the answer is correct if it is among the k first candidates[17].

Previous work on retrieval based dialogue modelling reports the following results for 1 in 10 R@1 score: Ubuntu 0.865 [13], Douban 0.259 [18], E-commerce Corpus 0.672 [11]. Furthermore, for the small, 500-sample Advisor dataset (DSTC7 Task 1) of choosing 10 responses from 100 (i.e., chance level also 0.1), accuracy of 0.630 was reported [12]. All above results were obtained with deep neural networks. While these results provide certain point of reference what accuracies one can expect for retrieval tasks, they are not directly comparable between each other or the result in this work. Not only is the language different, also sample sizes, preprocessing, dialogue themes and aims are varying. To the best of our knowledge, no comparable results have been reported on the response selection task for a less widely spoken language, such as Finnish.

3 Experimental Setting

Here we study multi-turn and single-turn chat data. In our multi-turn chat data, each dialogue contains one or more utterances from two conversationalists who alternate in turns. The last utterance is called a *response* and all utterance before that a *context* with one or more utterances. Context contains the relevant history for the response. The single-turn chat has only one utterance per person and is therefore often called as question-answering (QA) task. In our corpora each dialogue contains only two conversationalists.

Our task was to create retrieval-based models that were able to pick the best candidate response given a context (with one or more utterances). In contrast to generative models, testing the performance of a retrieval-based model is straightforward using 1 in 10 R@1 score. The response is always written by an expert, which is particularly important with the medical chat (i.e., the response is medically consistent) and no manual evaluation is mandatory in estimating model accuracy. The practical application we had in mind here was a response recommendation system that could provide off-the-shelf response candidates (e.g., top-5) for experts working with chat.

3.1 Description of the Data Sets

Medical Chat: This chat data was collected during 2016–2017. Goal of conversations was to solve a medical issue of the customer. The data was anonymized with all personal information censored (replaced with a special token), such as person names, social security numbers, addresses, emails and phone numbers. Only after anonymization was the data provided for us as JSON dumps. Each valid dialogue consisted of one or more utterances made by a customer and a medical professional. The customer was either a lay-person or another professional seeking consultation. The responder was always the expert. Being a highly sensitive dataset, it is not available publicly.

Library QA: This data was scraped from https://www.kirjastot.fi/kysy in April 2020 using a custom HTML crawler. Each dialogue included one question by a customer and the response by a librarian. Questions were open-domain and responses compact, fact sheet type of explanations often containing author and book recommendations and URLs to external sources. This data is fully public.

In both corpora the same person(s) could be present in multiple dialogues. In particular, one expert was often participating in lots of dialogues as it was part of their job duties. However, as no identification data of conversationalists was present, no exact statistics of this aspect was available.

In Table 1 we list some key descriptive statistics for both datasets. Although the datasets were similar with respect to token count (4.5M vs. 4.8M), many other characteristics were different. In particular, the vocabulary (i.e., number of unique tokens) of library QA data was twice that of the medical chat, however the ratio of the rare tokens (i.e., less than 5 occurrences in corpus) was similar (0.832 vs. 0.857). Library QA data had over 4 times the number of proper nouns compared to medical chat data. Also the responses by librarians were notably longer than those of medical professionals (11.8 vs. 86.0 tokens). These statistics were computed for uncased text after parsing (see Data preprocessing).

Table 1. Descriptive statistics of the two corpora used in this study. All characters were uncased before computation of statistics.

Property	Medical chat	Library QA
Tokens	4,460,124	4,835,696
Vocabulary size	177,720 (raw)	450,895 (raw)
	100,085 (lemma)	270,856 (lemma)
Rare token ratio	0.832 (raw)	0.857 (raw)
	0.838 (lemma)	0.846 (lemma)
Mean turns	10.83	2 (fixed)
Proper noun ratio	0.025	0.116
Number ratio	0.022	0.031
Punctuation ratio	0.165	0.189
Mean customer utterance tokens	16.66	27.50
Mean responder utterance tokens	11.82	86.02
Mean token length	5.21	6.25

3.2   Implementation and Parameters of Methods

We chose four methods to build retrieval-chatbot models: TF-IDF similarity, StarSpace, ESIM and BERT. These represent varying techniques ranging from the simple, classical ones (TF-IDF) to shallow neural networks (StarSpace), and from deep neural networks trained from scratch (ESIM) to massive pretrained networks (BERT). In addition to these, we also applied word2vec and fastText in other parts of analyses. For the medical chat task where the context usually contained multiple turns, all utterances were concatenated together and considered as long token vectors. Special end-of-utterance and end-of-turn tokens were added to inform models about the utterance and turn locations. In the following, we list technical details and parameter choices of the methods. If not mentioned, the remaining parameters were the defaults in the corresponding libraries at the time being (March 2020). Retrieval modelling was done using the ParlAI[3] framework [20].

TF-IDF Similarity: We used *TfidfVectorizer* from the scikit-learn library[4]. First we concatenated all dialogues (context and response) in the training data to compute vocabulary and weights in TF-IDF transformation. We took uni and bi-grams with minimum term count of 5. For inverse document frequency we used smoothing of 1 document (see, e.g., [17]). Then, for the test data, the context and response were separately vectorized and compared against each other via cosine similarity. The pair with the highest similarity over all candidates was considered as the correct match.

Word2vec, fastText and StarSpace: We used word2vec[5] to initialize embeddings for the ESIM, which was found better than training embeddings from scratch in ESIM. For this we used dimension 200, window length of 5 tokens, minimum word count 5 and CBOW algorithm [19]. fastText[6] was used to classify speaker roles in medical chat (see Data preprocessing). Parameters for the final model were: Learning rate 0.6, n-grams 1-2-3 with minimum term count of 10, char n-gram length between 2 and 15, embedding dimension 200 and context window of 5 terms. Finally, for StarSpace[7] the main parameters were: Learning rate 0.13, TF-IDF scaling on, negative samples 10 and embedding dimension of 200 with random initialization (using pretrained ones did not improve accuracy).

ESIM: We used the model implementation from https://github.com/jdongca2003/next_utterance_selection, where we added support for non-equal context and response truncation lengths. For the main parameters we set word embedding dimensions of 200, LSTM cell count 200, batch size 64 and maximum characters per word 20. Models were trained until no improvement for the development set during the last 3 epochs. Inputs were truncated into 450/100 (context) and 60/300 (response) tokens for medical/library tasks. These limits were set such that they covered over 95% of the training samples, hence truncation affected only the longest samples. Truncation was necessary to ensure that models fitted into the available GPU memory (here 32 GB).

---

[3] https://github.com/facebookresearch/ParlAI.
[4] https://github.com/scikit-learn/scikit-learn.
[5] https://github.com/tmikolov/word2vec.
[6] https://github.com/facebookresearch/fastText.
[7] https://github.com/facebookresearch/StarSpace.

BERT: We applied the BERT model for ranking as described in [13]. We used the simplest and fastest, Bi-encoder ranker model with aggregation of the first output layer as implemented in ParlAI framework. We obtained two pre-trained BERT models: Multilingual BERT[8] and FinBERT[9]. For both models, we obtained cased and uncased versions. Special tags used with the medical chat data (see below) were added into the model dictionary as new tokens. All models contained 12 layers with 12 attention heads with output dimension of 768. Input data was truncated similar to ESIM model. Models were trained with batch size of 32 until no improvement for the development set during the last 3 epochs.

3.3 Data Preprocessing

Before modelling, both corpora were preprocessed as described next. As a part of preprocessing, we applied a neural parser pipeline[10] to segment, morphologically tag (CoNLL-U format) and lemmatize texts [15]. We created four different versions of datasets for both tasks: Raw, lemmatized, spell-checked and lemmatized + spell-checked.

Lemmatization reduces vocabulary size by removing inflected forms of words. It is a standard preprocessing technique for classical, non-neural network methods, such linear text classification systems, but is rarely used with current neural network methods [5]. For the spell-checking, we used Voikko[11] tool, which we applied to rare words occurring in less than 5 samples in the training set. These words were then replaced with Voikko-suggested words (if any) and the same corrections were applied to development and test data, thus reducing the vocabulary size. We also applied both lemmatization and spell-checking by running the spell-checking for lemmatized words.

Medical Chat Data. Preprocessing of the medical chat data included three phases: (1) Converting raw JSON dumps into dialogues and running the parser, (2) rule-based speaker role identification and fastText labeling, and (3) refining, splitting and sub-sampling of dialogues. Dialogues were split between training, development and testing. In the following, the three phases are described in more detail.

In phase 1 we dropped all dialogues that contained less than 5 utterances, had only one participant, were not in Finnish and were duplicates. We were left with 29,602 dialogues. After this, all dialogues were parsed. As the raw data came unlabeled, i.e., no roles of conversationalists were specified, we came up with 42 commonly used, idiosyncratic phrases of the medical experts (responders) which were unlikely spoken by customers. For example, these included phrases (translated from Finnish) such as "how can I help you", "could I help you", "what do you need help with", "I will write a prescription" and their variations.

---

[8] https://huggingface.co/transformers/multilingual.html.
[9] https://github.com/TurkuNLP/FinBERT.
[10] https://turkunlp.org/Turku-neural-parser-pipeline.
[11] https://github.com/voikko/corevoikko.

If present in a dialogue, the dialogue was considered *complete* (labeled). If no such phrase was present in a dialogue OR was found for both conversationalists, the dialog was considered *incomplete* (10.3% of all). Finally we replaced numbers, proper nouns and URLs (if present) with special tags.

In phase 2, we trained a fastText model to classify roles (expert or customer) and used pseudo-labelling to increase our labeled data size. The model was trained using the rule-based labeled data (as discussed above) split in ratio of 90:5:5 between training, development and testing. We trained multiple models by varying key parameters and chose the model that maximized accuracy on development set. The best model with learning rate 0.6, containing 1-2-3-grams, minimum word count 10 and embedding dimension 200, reached F1 score of 0.998 for the development set and perfect score (1.000) for the test set. Those unlabeled dialogues where both roles were predicted with a probability of at least 0.9 were added into labeled data (84.8% of 3051), bringing the final labeled dialogue count into 29,139.

In the last, phase 3, we further dropped dialogues with less than 4 turns (with concatenated utterances; 1325 dialogues) or very high imbalance of utterance lengths (over 80% by a single conversationalist; 36 dialogues). The dataset was then split into training, development and testing sets using ratio 8:1:1. Finally, the three datasets were sub-sampled by extracting all sub-dialogues from the complete ones. For example, if a complete dialogue contained 10 turns, new samples were created with only 8,6 and 4 turns where the last is always by a medical expert. This allowed both full utilization of the data and also supported our aim of modelling all stages of a dialogue, from start ("welcome") to finish ("goodbye"). As we had to truncate the context size after 450 tokens, we kept up to 20 turns in modelling (the mean was 10.83, see Table 1).

Library QA Data. Unlike the medical chat data, only minimal preprocessing was required for the library data. After scraping and removing duplicate entries, total 42,552 question-answer pairs remained. Preprocessing included parsing and splitting into training, development and testing samples with ratio 8:1:1. Unlike medical chat, this data contained more URLs, which were truncated to first 20 characters, but otherwise kept intact. Here we also considered cased data, since this data was more formal and capitalization was systematic. This allowed us to test whether casing is actually useful in modelling (see, e.g., [29] for similar tests).

Preprocessing of data is illustrated in Fig. 1 for both tasks.

4   Results

In the following, we report the results of the modelling experiments. As a measure of model goodness we used 1 in 10 R@1 accuracy and values above 0.10 are considered better than chance. Training, development and testing samples were kept same for all methods. A new model was trained for each of the four types of datasets (i.e., raw, lemmatized, spell-checked and lemmatized + spell-checked)

(a)

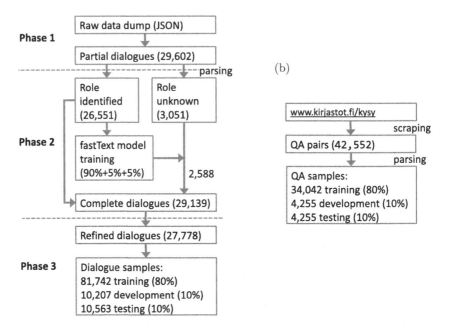

(b)

Fig. 1. Preprocessing pipeline for the (a) medical chat and (b) library QA datasets.

and for library QA task also for both cased and uncased data. The reported results are the best ones achieved for each model during testing.

Table 2. Accuracies for the medical chat task. Data and models were uncased. The last column shows the arithmetic mean over the other four columns. SC = Spell-Corrected.

Method	Raw	Lemma	SC	Lemma+SC	Mean
TF-IDF similarity	0.302	0.350	0.303	0.350	0.326
StarSpace	0.555	0.614	0.555	0.617	0.585
ESIM	0.798	0.816	0.805	0.818	0.809
ML BERT	0.859	0.842	0.860	0.860	0.855
FinBERT	0.921	0.896	0.922	0.881	0.905

Table 2 lists results for the medical chat task. The ranking of the methods is unambiguous with BERT models taking the lead, followed by ESIM, StarSpace and TF-IDF. FinBERT surpassed ML BERT by 5% (on average). The difference between ESIM and StarSpace was 22%, indicating a major gap between deep and shallow neural models.

Figure 2 depicts accuracy of ESIM and FinBERT models as a function of available turns in context (history). More turns allow the model more conversation history to make informed decision when choosing a correct response. The two models here are the ones trained with spell-checked data (4th column in Table 2). Accuracies start at 0.49 (ESIM) and 0.61 (FinBERT) and rapidly reach the plateau after 10 turns. Both ESIM and FinBERT exhibit similar behavior with respect to history size and results are similar to those previously reported for the Ubuntu and Douban tasks [18, 28].

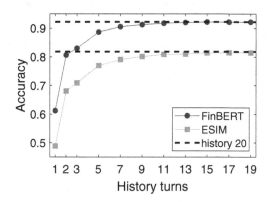

Fig. 2. Prediction accuracy for the medical chat as a function of maximum history turns for the best FinBert and ESIM models (see Table 2). Dashed lines depicts accuracies with all 20 turns used in training the models.

Table 3 lists results for the library QA task. All accuracies are higher than those of the medical chat, even the worst one reaching 0.599 for the uncased data. Again, FinBERT is the best model reaching near perfect score 0.987 (raw and uncased data). TF-IDF similarity was marginally better than StarSpace (by 5% on average).

Table 3. Accuracies for the library QA task. The last column shows the arithmetic mean over other four columns. SC = Spell-Corrected.

Method	Casing	Raw	Lemma	SC	Lemma+SC	Mean
TF-IDF similarity	Cased	0.637	0.803	0.640	0.807	0.722
	Uncased	0.677	0.815	0.678	0.816	0.746
StarSpace	Cased	0.493	0.772	0.590	0.768	0.656
	Uncased	0.599	0.792	0.652	0.788	0.708
ESIM	Cased	0.905	0.932	0.913	0.930	0.920
	Uncased	0.924	0.936	0.930	0.936	0.932
ML BERT	Cased	0.954	0.961	0.966	0.929	0.952
	Uncased	0.970	0.969	0.946	0.951	0.959
FinBERT	Cased	0.962	0.952	0.984	0.956	0.963
	Uncased	0.987	0.972	0.982	0.967	0.977

4.1 Output Examples

Finally, we present examples of the responses returned by our FinBERT models for medical and library tasks, both trained with spell-checked and uncased data. The example questions were made up by a person not involved in the study, hence they were not present in our corpora. We show top-3 responses and their ranking scores (higher is better). Questions and answers were translated into English, but other than that responses were kept close to their original form.

Medical Chat. Total number of unique responses over training, development and test sets was 82,995 meaning that 19% of all responses were repeated at least once. These were typically responses related to saying phrases like welcome, thanks or goodbye, or simple responses related, e.g., to prescriptions and asking about symptoms. Below are the top-3 responses picked among unique responses. Note that here we only consider one turn for the sake of simplicity, but one could continue the process over multiple turns by each time obtaining a new set of top-3 candidate responses.

Question 1: i feel sick, have a bad headache and temperature of <NUM> degrees. I feel very weak. What should i do?
response 1 (score 113.8): is headache your worst symptom?
response 2 (score 112.5): have you used any pain killers now?
response 3 (score 112.5): have you vomited?

Question 2: i feel stressed and anguished, and also have problems in sleeping. can you write me a prescription for some drug?
response 1 (score 108.4): what medications you have tried previously
response 2 (score 106.4): is there any other medicine than you mentioned
response 3 (score 106.2): certainly yes. i can give you some via electronic prescription. do you have any ongoing regular medication?

Library QA. Total number of unique responses over training, development and test sets was 42,468 meaning that only 0.2% of responses were not unique. Below are the top-3 responses picked among the unique responses. Long responses were clipped to save space (indicated by ". . . " symbol).

Question 1: i am interested in detective stories and thrillers. can you recommend me good books or suitable authors?
response 1 (score 124.7): here are few book tips from detective story writers:
camilla sten & viveca sten: synkät vedet -series (syvyyksissä, sumussa, loimut)
ilkka remes: piraatit, draculan ratsu, pimeän pyöveli, musta kobra

ari paulow: aavesaaren arvoitus, surmankarin vangit ...
response 2 (score 119.5): i recommend detective stories by karin fossum
response 3 (score 119.2): you can find rich selection of detective story links from makupalat-listing of the library of hämeenlinna at http://www. makupalat particularly in dekkarinetti- (http://www.tornio.fi) and suomen dekkariseura site (http://www.dekkarise) are worth visiting ...

Question 2: i am interested in 1800 - century handicraft and clothing. what books would you recommend for me to read?
response 1 (score 102.3): helmet-libraries have lots of good books about ryas, here are some
- rakas ryijy: suomalaisten ryijyt / pirkko sihvo
- ryijy esillä ryijyt suomen käsityön museon kokoelmissa / [edited by marjo kolu ... et al.]
- ryijy elää: suomalaisia ryijyjä 1778–2008 ...
response 2 (score 101.7): next books contain information about tapestries
grönroos, anne-maj: taidetekstiilien kutominen (class 65.41)
harjumäki, ulla: kankaankutojan ammattitekniikka (65.4107)
isotaito: otavan iso taitosanakirja, part 3, starting from page 223 (65.03)
...
response 3 (score 101.1): it's worth checking the following books: lehtinen, ildikó: rahwaan puku: näkökulmia suomen kansallismuseon kansanpukukokoelmiin
well heeled: the story of the finnish shoe suomalaisen arjen historia -series, parts 2 ja 3 (about everyday life) ...

Candidate scores for all available unique responses for both corpora are depicted in Fig. 3, sorted in decreasing order starting from rank 1. These represent the typical behavior of ranking scores.

(a)   (b)

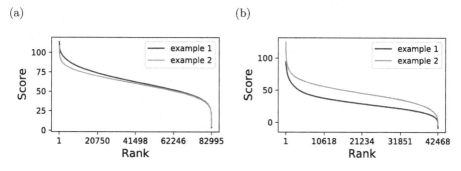

Fig. 3. Response scores for example questions using FinBERT for (a) medical chat and (b) library QA tasks. The higher score indicates better matching between a question and the response.

From above examples one can notice that all responses were on topic and most were also usable as is. However, as the pool of available responses for a retrieval-based model is fixed and limited, the responses are not always exact match with the question.

5 Analysis and Discussion

We analyzed two conversational corpora in Finnish: A public library question-answering data and a non-public medical chat data. We evaluated different retrieval models and preprocessing strategies for the response ranking task. Furthermore, we reviewed the best models by testing the effect of conversation history size and studied example predictions.

BERT Takes the Lead. For both tasks, best performance was found using BERT models. BERT leverages transfer learning technique by taking advantage of pretraining with massive unlabeled datasets [8,29]. Our fine-tuned Fin-BERT surpassed corresponding multilingual BERT by having up to 5% higher mean accuracy. This result was expected on the basis of recent findings on other Finnish language tasks [29]. The best results were found with uncased raw or spell-checked data. Although BERT worked well for all preprocessing types, lemmatization had generally negative impact on accuracy.

The second-best model was ESIM, which was trained from scratch. Mean accuracies of this model were 10% (medical chat) and 5% (library QA) lower than those of FinBERT, but still notably higher (18% or more) compared to non-deep neural network approaches. ESIM model has less parameters and it's computationally more reachable than BERT and remains a valid method if no suitable pretrained BERT model is available for finetuning. Unlike BERT, best results were achieved with lemmatized data.

Classic TF-IDF similarity method was the worst for the medical chat task (on average 26% behind StarSpace), but slightly surpassed (by 4%) a technically more advanced StarSpace model for the library QA task. The rich, multi-domain vocabulary of the library QA task makes a simple word-matching approach powerful in retrieval. On the other hand, when the vocabulary was impoverished and restricted to medical domain, the StarSpace model won TF-IDF with a notable margin. This result further emphasizes the importance of the task and corpus over the model selection. Although the performance of StarSpace was worse than those of the deep networks (ESIM and BERT), it has other advantages. Firstly, the method only requires modest computational resources; it easily parallelizes over CPUs (no GPU required) and has modest memory requirements. Finally, since TF-IDF and StarSpace are based on n-grams, they can handle documents of all lengths without need to truncate inputs. This is important if one needs to work with long texts.

Although our datasets were small compared to Ubuntu, Douban and E-commerce datasets, our top models achieved better accuracies. This could result from the fact that we used sub-sampling for medical chat task, where each dialogue was split into multiple samples (but never between training, development and test sets). This increases homogeneity of samples and could impoverish vocabulary when compared to a case where each sample comes from an individual dialogue. Smaller relative vocabulary can help in training. In addition to that, medical chat included lots of repeating responses (19%) which can be easier to predict than more complex ones. Finally, our preprocessing pipelines were carefully designed for both datasets and aimed to help our modelling efforts. However, as noted earlier, comparison to previous results is difficult due to differences in language, sample size, preprocessing and details of the task.

The Impact of Preprocessing. Classically, lemmatization can help in modelling by reducing the vocabulary size (see, e.g., [17]). This is particularly true for Finnish which has a rich morphology [15]. We found lemmatization generally useful for all but BERT models. In addition to having seen lots of words in pretraining, BERT also utilizes word pieces to handle out-of-vocabulary words [8].

Spell-checking increased accuracy in most of the test cases and was found generally useful for all models. On the other hand, the combination of lemmatization and spell-checking was found less useful or even harmful. The overall effect of spell-checking compared to raw data was often negligible (one percent or less) and might not be worth of the extra computational effort. Spell-checking can also introduce new mistakes in the data, which is especially true for domains with specialized and rare vocabulary (e.g., medicine).

For the library QA task, we also tested the effect of casing and the best results were found with uncased data. Only 2 out of 8 BERT models reached marginally better results (up to 2%) with casing.

Inference Speed and Model Updating. Apart from accuracy, the satisfactory inference speed is also essential for deployment of a ranking model. All four models tested here allow precomputing of candidate responses such that the inference can be reduced into computing dot products pf vectors at runtime. With a high-performance system with modern GPUs, this allows ranking of tens of thousands of candidates in just a matter of milliseconds (see, e.g., [13]). Another important aspect of a production system is regular updating (re-training) of the model with new data. New data can include both new dialogues, but also user feedback (e.g., suggested corrections) for model suggested responses, thus allowing improvement of the model over time.

Limitations of the Work. This study evaluated four model types, but this is only a small sample of all available model types. In particular, there are various deep neural network models that are expected to reach accuracies close to ESIM (see, e.g., [13,28]) and also different variations of BERT for ranking [13]. Our experiments were performed using mostly default parameters. No systematic global optimization of the parameters was performed as it was considered computationally infeasible. In this work we concentrated on testing model accuracy without further development into optimized, production -ready systems. For inference speed tests and related discussion that also apply to this work (particularly BERT models), one can check Ref. [13]. Finally, our conclusions related to model performance and preprocessing are based on our two datasets and all conclusions might not generalize to different languages and datasets. As evident based on this work, optimal preprocessing and model selection depend strongly on the data.

6   Conclusions

BERT models were found the best in all cases tested in this study, thus highlighting the power of the pretrained language models. The model pretrained for Finnish (FinBERT) was the overall best. Preprocessing had less impact for BERT models compared to the classical (TF-IDF) and other neural network models (ESIM and StarSpace). We found that the classical TF-IDF method was a strong baseline for the vocabulary-rich library QA task, even surpassing StarSpace. The best accuracies were reached using uncased text with spell-checking (BERT models) or lemmatization (non-BERT models).

The accuracy in the response ranking task of our best models for both datasets was better than the results obtained for the well known one-to-one dialogue corpora in the English and Chinese languages. To our knowledge, this is the first work on dialogue modelling using any type of neural networks for the Finnish language. Previous work on dialogue modelling in any other European language is scant and comparable results have not been reported. Our work contributes towards the development of automated dialogue systems, also called goal-oriented chatbots, in the professional domain.

The next step in modelling would be to expand into generative models based on FinBERT. This would allow responses to such contexts that currently do not have suitable existing responses in the dataset. In this effort, manual response evaluation becomes important, as evaluation cannot be done simply by taking apart a separate test dataset from the original dialogue corpus. However, this effort might be worthwhile as generative models appear promising in a setting where the dialogues evolve over time and new topics and ways of expression appear.

Acknowledgement. The authors wish to acknowledge CSC – IT Center for Science, Finland, for computational resources.

References

1. The suomi24 sentences corpus 2001–2017, korp version 1.1. http://urn.fi/urn:nbn: fi:lb-2020021803

2. Aunimo, L., Heinonen, O., Kuuskoski, R., Makkonen, J., Petit, R., Virtanen, O.: Question answering system for incomplete and noisy data. In: Sebastiani, F. (ed.) ECIR 2003. LNCS, vol. 2633, pp. 193–206. Springer, Heidelberg (2003). https://doi.org/10.1007/3-540-36618-0_14

3. Beißwenger, M., Bartsch, S., Evert, S., Würzner, K.M.: EmpiriST 2015: a shared task on the automatic linguistic annotation of computer-mediated communication and web corpora. In: Proceedings of the 10th Web as Corpus Workshop, pp. 44–56 (2016)

4. Beißwenger, M., Storrer, A.: 21. Corpora of computer-mediated communication. Corpus Linguistics. An International Handbook. Series: Handbücher zur Sprach- und Kommunikationswissenschaft/Handbooks of Linguistics and Communication Science. Mouton de Gruyter, Berlin (2008)

5. Camacho-Collados, J., Pilehvar, M.T.: On the role of text preprocessing in neural network architectures: an evaluation study on text categorization and sentiment analysis. In: Proceedings of the 2018 EMNLP Workshop BlackboxNLP: Analyzing and Interpreting Neural Networks for NLP, Stroudsburg, PA, USA, pp. 40–46. Association for Computational Linguistics (2018). https://doi.org/10.18653/v1/W18-5406

6. Chaudhuri, D., Kristiadi, A., Lehmann, J., Fischer, A.: Improving response selection in multi-turn dialogue systems by incorporating domain knowledge. In: Proceedings of the 22nd Conference on Computational Natural Language Learning, Brussels, Belgium. Association for Computational Linguistics (2018)

7. Chen, Q., Zhu, X., Ling, Z.H., Wei, S., Jiang, H., Inkpen, D.: Enhanced LSTM for natural language inference. In: Proceedings of the 55th Annual Meeting of the Association for Computational Linguistics (Volume 1: Long Papers), Vancouver, Canada, pp. 1657–1668. Association for Computational Linguistics (2017). https://doi.org/10.18653/v1/P17-1152

8. Devlin, J., Chang, M.W., Lee, K., Toutanova, K.: BERT: pre-training of deep bidirectional transformers for language understanding. In: Proceedings of the 2019 Conference of the North American Chapter of the Association for Computational Linguistics: Human Language Technologies (Volume 1: Long and Short Papers), Minneapolis, Minnesota, pp. 4171–4186. Association for Computational Linguistics (2019). https://doi.org/10.18653/v1/N19-1423

9. Dong, J., Huang, J.: Enhance word representation for out-of-vocabulary on ubuntu dialogue corpus. arXiv preprint arXiv:1802.02614 (2018)

10. Gu, J.C., Li, T., Liu, Q., Zhu, X., Ling, Z.H., Ruan, Y.P.: Pre-trained and attention-based neural networks for building noetic task-oriented dialogue systems. arXiv preprint arXiv:2004.01940 (2020)

11. Gu, J.C., Ling, Z.H., Liu, Q.: Interactive matching network for multi-turn response selection in retrieval-based chatbots. In: Proceedings of the 28th ACM International Conference on Information and Knowledge Management, CIKM 2019, New York, NY, USA, pp. 2321–2324. Association for Computing Machinery (2019). https://doi.org/10.1145/3357384.3358140

12. Gunasekara, C., Kummerfeld, J.K., Polymenakos, L., Lasecki, W.: DSTC7 task 1: noetic end-to-end response selection. In: Proceedings of the First Workshop on NLP for Conversational AI, Florence, Italy, pp. 60–67. Association for Computational Linguistics (2019). https://doi.org/10.18653/v1/W19-4107

13. Humeau, S., Shuster, K., Lachaux, M.A., Weston, J.: Poly-encoders: transformer architectures and pre-training strategies for fast and accurate multi-sentence scoring. arXiv preprint arXiv:1905.01969 (2019)

14. Joulin, A., Grave, E., Bojanowski, P., Mikolov, T.: Bag of tricks for efficient text classification. In: Proceedings of the 15th Conference of the European Chapter of the Association for Computational Linguistics: Volume 2, Short Papers, pp. 427–431. Association for Computational Linguistics (2017)
15. Kanerva, J., Ginter, F., Miekka, N., Leino, A., Salakoski, T.: Turku neural parser pipeline: an end-to-end system for the CoNLL 2018 shared task. In: Proceedings of the CoNLL 2018 Shared Task: Multilingual Parsing from Raw Text to Universal Dependencies. Association for Computational Linguistics (2018)
16. Lowe, R., Pow, N., Serban, I.V., Pineau, J.: The ubuntu dialogue corpus: a large dataset for research in unstructured multi-turn dialogue systems. In: Proceedings of the 16th Annual Meeting of the Special Interest Group on Discourse and Dialogue, pp. 285–294 (2015)
17. Manning, C.D., Raghavan, P., Schütze, H.: An Introduction to Information Retrieval. Cambridge University Press, Cambridge (2009)
18. Mao, G., Jindian, S., Yu, S., Luo, D.: Multi-turn response selection for chatbots with hierarchical aggregation network of multi-representation. IEEE Access **PP**, 1 (2019). https://doi.org/10.1109/ACCESS.2019.2934149
19. Mikolov, T., Sutskever, I., Chen, K., Corrado, G., Dean, J.: Distributed representations of words and phrases and their compositionality. In: Proceedings of the 26th International Conference on Neural Information Processing Systems, NIPS 2013, vol. 2, pp. 3111–3119. Curran Associates Inc., Red Hook (2013)
20. Miller, A., et al.: ParlAI: a dialog research software platform. In: Proceedings of the 2017 Conference on Empirical Methods in Natural Language Processing: System Demonstrations, pp. 79–84 (2017)
21. Numminen, P.: Kysy kirjastonhoitajalta-neuvontapalvelun kysymystyypit. Informaatiotutkimus **27**(2), 55–60 (2008)
22. Riou, M., Salim, S., Hernandez, N.: Using discursive information to disentangle French language chat (2015)
23. Ritter, A., Cherry, C., Dolan, B.: Unsupervised modeling of twitter conversations. In: Human Language Technologies: The 2010 Annual Conference of the North American Chapter of the Association for Computational Linguistics, pp. 172–180. Association for Computational Linguistics (2010)
24. Serban, I.V., Lowe, R., Henderson, P., Charlin, L., Pineau, J.: A survey of available corpora for building data-driven dialogue systems. Dialogue Discourse **9**(1), 1–49 (2018)
25. Shang, L., Lu, Z., Li, H.: Neural responding machine for short-text conversation. In: Proceedings of the 53rd Annual Meeting of the Association for Computational Linguistics and the 7th International Joint Conference on Natural Language Processing (Volume 1: Long Papers), pp. 1577–1586 (2015)
26. Swanson, K., Yu, L., Fox, C., Wohlwend, J., Lei, T.: Building a production model for retrieval-based chatbots. In: Proceedings of the First Workshop on NLP for Conversational AI, pp. 32–41 (2019)
27. Vaswani, A., et al.: Attention is all you need. In: Guyon, I., et al. (eds.) Advances in Neural Information Processing Systems 30, pp. 5998–6008. Curran Associates, Inc. (2017)
28. Vig, J., Ramea, K.: Comparison of transfer-learning approaches for response selection in multi-turn conversations. In: Association for the Advancement of Artificial Intelligence (2019)
29. Virtanen, A., et al.: Multilingual is not enough: BERT for Finnish. arXiv preprint arXiv:1912.07076 (2019)

30. Wang, H., Lu, Z., Li, H., Chen, E.: A dataset for research on short-text conversations. In: Proceedings of the 2013 Conference on Empirical Methods in Natural Language Processing, pp. 935–945 (2013)
31. Wu, L.Y., Fisch, A., Chopra, S., Adams, K., Bordes, A., Weston, J.: StarSpace: embed all the things! In: Thirty-Second AAAI Conference on Artificial Intelligence (2018)
32. Wu, Y., Wu, W., Xing, C., Xu, C., Li, Z., Zhou, M.: A sequential matching framework for multi-turn response selection in retrieval-based chatbots. Comput. Linguist. **45**(1), 163–197 (2019)
33. Wu, Y., Wu, W., Xing, C., Zhou, M., Li, Z.: Sequential matching network: a new architecture for multi-turn response selection in retrieval-based chatbots. In: Proceedings of the 55th Annual Meeting of the Association for Computational Linguistics (Volume 1: Long Papers), pp. 496–505 (2017)

Advances of Transformer-Based Models for News Headline Generation

Alexey Bukhtiyarov$^{(\boxtimes)}$ and Ilya Gusev

Moscow Institute of Physics and Technology, Moscow, Russia
{bukhtiyarov.ao,ilya.gusev}@phystech.edu

Abstract. Pretrained language models based on Transformer architecture are the reason for recent breakthroughs in many areas of NLP, including sentiment analysis, question answering, named entity recognition. Headline generation is a special kind of text summarization task. Models need to have strong natural language understanding that goes beyond the meaning of individual words and sentences and an ability to distinguish essential information to succeed in it. In this paper, we fine-tune two pretrained Transformer-based models (mBART and Bert-SumAbs) for that task and achieve new state-of-the-art results on the RIA and Lenta datasets of Russian news. BertSumAbs increases ROUGE on average by 2.9 and 2.0 points respectively over previous best score achieved by Phrase-Based Attentional Transformer and CopyNet.

Keywords: Text summarization · Headline generation · Russian language · BERT

1 Introduction

Text summarization aims to condense vital information from text into a shorter, coherent form that includes main ideas. Two main approaches are distinguished: extractive, which involves organizing words and phrases extracted from text to create a summary, and abstractive, which requires the ability to generate novel phrases not featured in the source text while preserving the meaning and essential information.

Headline generation is considered within the automatic text generation area, so these methods are conventional approaches to that task. Because headlines are usually shorter than summaries, the model has to be good at distinguishing the most salient theme and compressing it in a syntactically correct way. The task is vital for news agencies and especially news aggregators [1]. The right solution can be beneficial both for them and for the readers. The headline is the most widely read part of any article, and due to its summarization abilities, it can help decide whether a particular article is worth spending time.

An essential property of the headline generation task is data abundance. It is much easier to collect a dataset in any language containing articles with

© Springer Nature Switzerland AG 2020
A. Filchenkov et al. (Eds.): AINL 2020, CCIS 1292, pp. 54–61, 2020.
https://doi.org/10.1007/978-3-030-59082-6_4

headlines than articles with summaries because articles usually have a headline by default. We can use the headline generation as a pretraining phase for other problems like text classification or clustering news articles. This two-stage fine-tuning approach is shown to be effective [2]. That is why it is essential to investigate the performance of different models on this particular task.

In this paper, we explore the effectiveness of applying the pretrained Transformer-based models for the task of headline generation. Concretely, we fine-tune mBART and BertSumAbs models and analyze their performance. We obtain results that validate the applicability of these models to the headline generation task.

2   Related Work

Previous advances in abstractive text summarization have been made using RNNs with an attention mechanism [3].

One more important technique to improve RNN encoder-decoder model is copying mechanism [15] that increases the model ability to copy tokens from the input. LSTM-based CopyNet on byte pair encoded tokens achieved the previous state-of-the-art results on Lenta dataset [12]. In the Pointer-Generator network (PGN) that idea was further developed by introducing coverage mechanism [16] to keep track of what has been summarized.

The emergence of pretrained models based on Transformer architecture [4] led to new improvements. Categorization, history and applications of these models are comprehensively described in the survey [5]. Applying a Phrase-Based Attentional Transformer (PBATrans) achieved the latest state-of-the-art results on RIA dataset that we are also considering [10].

3   Models Description

Applying Transformer-based models usually follows these steps. Firstly, during unsupervised training, the model learns the universal representation of language. Then it is fine-tuned to a downstream task. Below we briefly describe the pretrained models we focus on in this work.

The first model is mBART [6]. It is a standard Transformer-based model consisting of an encoder and autoregressive decoder. It is trained by reconstructing the document from a noisy version of that document. Document corruption strategies include randomly shuffling the original sentences' order, randomly rotating the document so that it starts from different token and text infilling schemes with different span lengths. mBART is pretrained once for all languages on a subset of 25 languages. It has 12 encoder layers and 12 decoder layers with a total of ∼680M parameters, and its vocabulary size is 250K.

The second model we examine is BertSumAbs [7]. It utilizes pretrained Bidirectional Encoder Representations from Transformers (BERT) [8] as a building block for the encoder. The encoder itself is a 6 stacked layers of BERT. Multi-sentence representations [CLS] tokens are added between sentences, and interval segmentation embeddings are used to distinguish multiple sentences within a document. The decoder is randomly initialized 6-layered Transformer. However, because the encoder is pretrained and the decoder is trained from scratch, the fine-tuning may be unstable. Following [7], we use a fine-tuning schedule that separates the optimizers to handle this mismatch.

As a pretrained BERT for BertSumAbs, we use RuBERT trained on the Russian part of Wikipedia and news data [9]. A vocabulary of Russian subtokens of 120K was built from this data as well. BertSumAbs has ~320M parameters.

Scripts and models' checkpoints for both models are available[1].

4   Datasets

The RIA dataset consists of news articles and associated headlines for around 1 million examples [11]. These documents were published on the website "RIA Novosti (RIA news)" from January 2010 to December 2014. For training purposes, we split the dataset into the train, validation, and test parts in a proportion of 90:5:5.

The Lenta dataset[2] contains about 800 thousand news articles with titles that were published from 1999 to 2018. The purpose of model evaluation on this dataset is to measure the model capabilities to generate summaries given articles with another structure, different period, and style of writing.

There are no timestamps in both datasets, so time-based splits are unavailable. It can cause some bias because models tend to perform better on texts with entities that they saw during training.

5   Experiments

5.1   Evaluation

To evaluate the model, we use the ROUGE metric [13], reporting F_1 score for unigram and bigram overlap (ROUGE-1, ROUGE-2), and the longest common subsequence (ROUGE-L). We use the mean of these three metrics as the primary metric (R-mean). To balance this with a precision-based measure, we report BLEU [14].

We also reporting the proportion of novel n-grams that appear in the summaries but not in the source texts. The higher that number is the fewer copying was made to generate a summary meaning more abstractive model.

As baseline models we use first sentence and PGN. First sentence is a strong baseline for news articles because often the most valuable information is at the start while further sentences provide details and background information. PGN on byte-pair encoded tokens approach was described in the paper [12]. The implementation and parameters for PGN are used from here[3].

---

[1]   https://github.com/leshanbog/PreSumm.
[2]   https://github.com/yutkin/Lenta.Ru-News-Dataset.
[3]   https://github.com/IlyaGusev/summarus.

5.2   Training Dynamics

In Fig. 1 we present the training dynamics of the BertSumAbs model. It takes about 3 days to train 45K steps (on GeForce GTX 1080) with a batch size equals 8, gradient accumulation every 95 steps. We use 40K checkpoint as a final model due to its better loss score on the validation dataset.

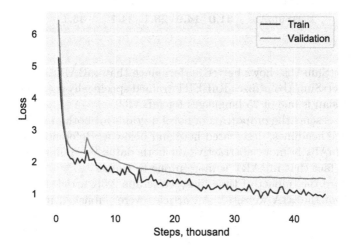

Fig. 1. BertSumAbs training dynamics on train and validation parts

Table 1. RIA dataset evaluation

Model	R1	R2	RL	R-mean	BLEU
First sentence	23.8	10.5	16.6	16.9	21.8
CopyNet [12]	41.6	24.5	38.9	35.0	53.8
PBATrans [10]	43.0	25.4	40.0	36.1	–
PGN	42.3	25.1	39.6	35.7	54.2
mBART	42.8	25.5	39.9	36.1	55.1
BertSumAbs	**46.0**	**28.0**	**43.1**	**39.0**	**57.6**

6   Results

Table 1 demonstrates the evaluation results on the RIA dataset. There is a significant improvement in all considered metrics for BertSumAbs whereas mBART performance is on the previous state-of-the-art level. Table 2 presents results on the Lenta dataset while models are trained on the RIA dataset. Both models show an improvement compared to previous results for all metrics. On both

Table 2. Lenta dataset evaluation using model trained on RIA dataset

Model	R1	R2	RL	R-mean	BLEU
First sentence	24.0	10.6	18.3	17.6	24.9
CopyNet [12]	28.3	14.0	25.8	22.7	40.4
PGN	26.4	12.3	24.0	20.9	39.8
mBART	30.3	14.5	27.1	24.0	43.2
BertSumAbs	**31.0**	**14.9**	**28.1**	**24.7**	**45.1**

datasets, BertSumAbs shows better performance than mBART. A possible reason is that BertSumAbs utilizes RuBERT trained specifically on Russian corpus, whereas Russian is one of 25 languages for mBART.

Figure 2 presents the proportion of novel n-grams for both models in comparison with true headlines, designated here and below as Reference. Results show that BertSumAbs is more abstractive on both datasets. A manual inspection confirms the fact that mBART is more prone to coping.

To measure the impact of time bias, we scrape 2.5K articles and associated headlines from the RIA website[4]. All articles were published in the 2020 year. The results are shown in Table 3. As expected, there is a decrease in metrics which is explained by new entities in articles. BertSumAbs suffers a greater decrease than mBART which may be because of more diverse pretraining of the latter.

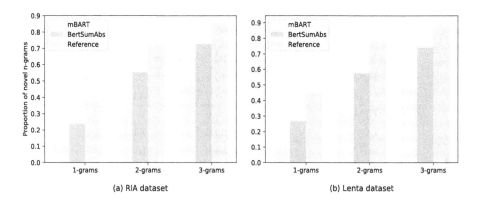

(a) RIA dataset   (b) Lenta dataset

Fig. 2. Proportion of novel n-grams in headlines

---

[4] https://ria.ru/.

Table 3. Evaluation on RIA articles from 2020

Model	R1	R2	RL	R-mean	BLEU
PGN	39.6	20.8	35.2	31.9	52.1
mBART	41.7	22.7	37.2	33.9	53.2
BertSumAbs	41.9	22.5	37.3	33.9	54.2

Table 4. Bad examples

1. Reference	курс доллара подрос на открытии в среду на 1 коп - до 28,04 руб en: *the dollar rate increased at the opening on wednesday by 1 kopeck - up to 28,04 rubles*
1. mBART	рубль укрепился на открытии на открытии на открытии на одну копейку en: *ruble strengthened at the opening at the opening at the opening by 1 kopeck*
1. BertSumAbs	курс доллара подрос на открытии в среду на 1 коп - до 28,04 руб en: *the dollar rate increased at the opening on wednesday by 1 kopeck - up to 28,04 rubles*
2. Reference	7 дней в гаване: неделя в кубе en: *7 days in havana: a week in cuba*
2. mBART	"экстранхерос" или "виахерос" en: *"extranheros" or "viheros"*
2. BertSumAbs	кубинская мама, или как таксисты застряли на таможне en: *cuban mom, or how taxi drivers got stuck at customs*
3. Reference	женщина без рук стирает ногами, а зубами развешивает белье en: *a woman without hands washes her feet and hangs clothes with her teeth*
3. mBART	безрукой женщине, живущей в калужской области, говорят: "имела бы ты руки" en: *to armless woman living in the kaluga region say: "would you have hands"*
3. BertSumAbs	безрукая женщина в калужской области родила безрукую женщину en: *armless woman in the kaluga region gave birth to an armless woman*
4. Reference	китаец четыре года жил с ножом в голове, не зная об этом en: *for four years a chinese man lived with a knife in his head, not knowing about it*
4. mBART	врачи обнаружили лезвие ножа в голове своего пациента en: *doctors found a knife blade in their patient's head*
4. BertSumAbs	врачи обнаружили в голове жителя китая десятисантиметровое ружье en: *doctors found in the head of a resident of china a ten-centimeter gun*

6.1 Human Evaluation

To understand model performance beyond automatic metrics, we conducted the human evaluation of the results. We randomly sample 1110 articles and show them both with true and predicted headlines to human annotators. We use Yandex.Toloka[5] to manage the process of distributing samples and recruiting annotators. The task was to decide which headline is better: generated by Bert-SumAbs or by human. There was also a draw option if the annotator could not decide which headline is better. There was 9 annotators for each example and they did not know about the origin of the headlines. As a result, we get draw in 8% and BertSumAbs win in 49% of samples. Analysing the aggregate statistics, we found that BertSumAbs headlines were chosen by 5 or more annotators in 32% whereas human generated in 28%.

6.2 Error Analysis

Although, in general, the model's output is fluent and grammatically correct, some common mistakes are intrinsic for both models. In Table 4, we provide several such examples. In the first example, there is a repeated phrase in mBART output, while BertSumAbs is very accurate. The second example is so hard for both models that there are imaginary words in mBART prediction. It is a common problem for models operating on a subword level. The third example confirms that some articles are hard for both models. One more type of mistake is factual errors, as in the fourth example, where BertSumAbs reported about the wrong subject. This type of error is the worst because it is the most difficult to detect.

7   Conclusion and Future Work

In this paper, we showcase the effectiveness of fine-tuning the pretrained Transformer-based models for the task of abstractive headline generation and achieve new state-of-the-art results on Russian datasets. We showed that BertSumAbs that utilize language-specific encoder has better results than mBART. Moreover, human evaluation confirms that BertSumAbs is capable of generating headlines indistinguishable from the human-created ones.

In future work, we should use the headline generation task as a first step in the two-stage fine-tuning strategy to transfer knowledge to other tasks, such as news clustering and classification.

References

1. Murao, K., et al.: A case study on neural headline generation for editing support. In: Proceedings of the 2019 Conference of the North American Chapter of the Association for Computational Linguistics: Human Language Technologies (Volume 2: Industry Papers), pp. 73–82 (2019)

---

[5] https://toloka.yandex.com/.

2. Sun, C., Qiu, X., Xu, Y., Huang, X.: How to fine-tune BERT for text classification? In: Sun, M., Huang, X., Ji, H., Liu, Z., Liu, Y. (eds.) CCL 2019. LNCS (LNAI), vol. 11856, pp. 194–206. Springer, Cham (2019). https://doi.org/10.1007/978-3-030-32381-3_16

3. Rush, A.M., Chopra, S., Weston, J.: A neural attention model for abstractive sentence summarization. In: Empirical Methods in Natural Language Processing, pp. 379–389 (2015)

4. Vaswani, A., et al.: Attention is all you need. In: NeurIPS (2017)

5. Qiu, X., Sun, T., Xu, Y., Shao, Y., Dai, N., Huang, X.: Pre-trained models for natural language processing: a survey (2020). arXiv preprint arXiv:2003.08271

6. Liu, Y., et al.: Multilingual denoising pre-training for neural machine translation (2020). arXiv preprint arXiv:2001.08210

7. Liu, Y., Lapata, M.: Text summarization with pretrained encoders. In: Proceedings of the 2019 Conference on Empirical Methods in Natural Language Processing and the 9th International Joint Conference on Natural Language Processing (EMNLP-IJCNLP) (2019)

8. Devlin, J., Chang, M.W., Lee, K., Toutanova, K.: BERT: pre-training of deep bidirectional transformers for language understanding (2018). arXiv preprint arXiv:1810.04805

9. Kuratov, Y., Arkhipov, M.: Adaptation of deep bidirectional multilingual transformers for Russian language (2019). arXiv preprint arXiv:1905.07213

10. Sokolov, A.: Phrase-based attentional transformer for headline generation. In: Computational Linguistics and Intellectual Technologies: Proceedings of the International Conference "Dialogue 2019" (2019)

11. Gavrilov, D., Kalaidin, P., Malykh, V.: Self-attentive model for headline generation. In: Azzopardi, L., Stein, B., Fuhr, N., Mayr, P., Hauff, C., Hiemstra, D. (eds.) ECIR 2019. LNCS, vol. 11438, pp. 87–93. Springer, Cham (2019). https://doi.org/10.1007/978-3-030-15719-7_11

12. Gusev, I.O.: Importance of copying mechanism for news headline generation. In: Komp'juternaja Lingvistika i Intellektual'nye Tehnologii, pp. 229–236. ABBYY Production LLC (2019)

13. Lin, C.Y., Och, F.J.: ROUGE and its evaluation. In: NTCIR Workshop, Looking for a Few Good Metrics (2004)

14. Papineni, K., Roukos, S., Ward, T., Zhu, W.J.: BLEU: a method for automatic evaluation of machine translation. In: 40th Annual meeting of the Association for Computational Linguistics, pp. 311–318 (2002)

15. Gu, J., Lu, Z., Li, H., Li, V.O.: Incorporating copying mechanism in sequence-to-sequence learning (2016). arXiv preprint arXiv:1603.06393

16. See, A., Liu, P., Manning, C.: Get to the point: summarization with pointer-generator networks. In: Proceedings of the 55th Annual Meeting of the Association for Computational Linguistics, Vancouver, vol. 1, pp. 1073–1083. Association for Computational Linguistics (2017)

An Explanation Method for Black-Box Machine Learning Survival Models Using the Chebyshev Distance

Lev V. Utkin$^{(\boxtimes)}$ ⓘ, Maxim S. Kovalev ⓘ, and Ernest M. Kasimov ⓘ

Peter the Great St. Petersburg Polytechnic University, Saint-Petersburg, Russia
lev.utkin@gmail.com, maxkovalev03@gmail.com, kasimov.ernest@gmail.com

Abstract. A new modification of the explanation method SurvLIME called SurvLIME-Inf for explaining machine learning survival models is proposed. The basic idea behind SurvLIME as well as SurvLIME-Inf is to apply the Cox proportional hazards model to approximate the black-box survival model at the local area around a test example. The Cox model is used due to the linear relationship of covariates. In contrast to SurvLIME, the proposed modification uses L_∞-norm for defining distances between approximating and approximated cumulative hazard functions. This leads to a simple linear programming problem for determining important features and for explaining the black-box model prediction. Moreover, SurvLIME-Inf outperforms SurvLIME when the training set is very small. Numerical experiments with synthetic and real datasets demonstrate the SurvLIME-Inf efficiency.

Keywords: Explainable AI · Survival analysis · Censored data · Cox model · Chebyshev distance

1 Introduction

Explainability of deep learning models is a topical direction of research nowadays, and, as a result, a lot of methods have been developed to address the interpretation problems and to get accurate explanations for obtained predictions [5, 8, 14–16].

There are two main groups of methods for explaining the black-box models: local methods which aim to interpret a single prediction, and global methods which explain a black-box model on the whole input space. We study only the local models because our aim is to find features which lead to the individual prediction. Moreover, the so-called post-hoc explanation methods are considered, which are used to explain predictions of black-box models after they are trained.

One of the well-known local post-hoc explanation methods is the Local Interpretable Model-agnostic Explanations (LIME) [18], which uses easily understandable linear models to locally approximate the predictions of black-box models. LIME provides an explanation for an instance by perturbing it around its

This work is supported by the Russian Science Foundation under grant 18-11-00078.

A. Filchenkov et al. (Eds.): AINL 2020, CCIS 1292, pp. 62–74, 2020.
https://doi.org/10.1007/978-3-030-59082-6_5

neighborhood and then by constructing a local linear model. Following LIME, a lot of its modifications have been developed, for example, DLIME [23], Anchor LIME [19], NormLIME [1]. We have to point out also other explanation methods, including the SHAP method [13], counterfactual explanations [21], perturbation techniques [7], and many others. Detailed descriptions of many explanation methods can be found in survey papers [2, 5, 8, 20].

One of the peculiarities of LIME as well as other explanation models is that they explain point-valued predictions produced by the black-box model. However, there exist models which produce functions as predictions instead of points, for example, machine learning survival models [22] which solve survival analysis tasks [10]. One of the most widely-used regression models for the analysis of survival data is the well-known Cox proportional hazards model, which calculates effects of observed covariates on the risk of an event occurring, for example, death or failure [6]. The model assumes that a patient's log-risk of failure is a linear combination of the instance covariates. This is a very important and too strong assumption. Therefore, there are many survival analysis models, for example, random survival forests (RSFs) [4], deep neural networks [12], etc., which relax this assumption and allow for more general relationships between covariates and the output parameters [17, 22]. However, these models are the black-box ones and, therefore, they require to be explained. Taking into account that predictions of the models are functions, for example, the survival function (SF), the cumulative hazard function (CHF), the original LIME or other methods cannot be directly used. Kovalev et al. [11] proposed an explanation method called SurvLIME, which deals with censored data. The basic idea behind SurvLIME is to apply the Cox model to approximate the black-box survival model at a local area around a test instance. The Cox model is chosen due to its assumption of the linear combination of covariates. Moreover, it is important that the covariates as well as their combination do not depend on time. Therefore, coefficients of the covariates can be regarded as quantitative impacts on the prediction.

SurvLIME includes a procedure which randomly generates synthetic instances around the tested instance, and the CHF is calculated for every synthetic instance by means of the black-box survival model. For every instance, the approximating Cox model CHF is written as a function of coefficients of interest. By writing the distance between CHFs provided by the black-box survival model and by the approximating Cox model, respectively, an unconstrained convex optimization problem for computing the coefficients of covariates is constructed. The L_2-norm is used in order to consider the distance between two CHF. As a result, the explanation by using SurvLIME is based on solving the convex optimization problem. In order to simplify the approach, we propose and investigate another explanation method which is based on using L_∞-norm for the distance between CHFs. This modification is called SurvLIME-Inf.

The choice of this norm is caused by the fact that obtained optimization problems become rather simple from the computational point of view. Indeed, we get the linear optimization problem for computing coefficients of the Cox model. The L_∞-norm (Chebychev distance) is a measure of the approximation quality, which is defined as the maximum of absolute values of the difference between the function being approximated and the approximating function. Our experiments using synthetic and real data have illustrated a perfect approximation of CHFs provided by the black-box survival model and the approximating Cox model by rather small datasets.

The paper is organized as follows. Basic concepts of survival analysis are considered in Sect. 2. A brief introduction to LIME can be found in Sect. 3. Basic ideas behind SurvLIME-Inf are proposed in Sect. 4. Section 5 contains a formal derivation of the linear programming problem implementing SurvLIME-Inf. Numerical experiments with synthetic and real data are given in Sect. 6. Concluding remarks are provided in Sect. 7.

2   Basic Definitions of Survival Analysis

In survival analysis, an instance (patient) i is represented by a triplet $(\mathbf{x}_i, \delta_i, T_i)$, where $\mathbf{x}_i^{\mathrm{T}} = (x_{i1}, ..., x_{id})$ is the feature vector; T_i is time to event of the instance. If an event of interest is observed, then $\delta_i = 1$ and T_i is the time between baseline time and the time of event happening (an uncensored observation). If the event is not observed and its time to event is greater than the observation time, then $\delta_i = 0$ and T_i is the time between baseline time and end of the observation (a censored observation). Suppose a training set D consists of n triplets $(\mathbf{x}_i, \delta_i, T_i)$, $i = 1, ..., n$. Survival analysis aims to estimate the time to the event of interest T for a new instance with a feature vector $\mathbf{x}$ by using D.

The SF $S(t|\mathbf{x})$ and CHF $H(t|\mathbf{x})$ are important concepts in survival analysis [10]. The SF is the probability of surviving up to time t, i.e., $S(t|\mathbf{x}) = \Pr\{T > t|\mathbf{x}\}$. The CHF $H(t|\mathbf{x})$ is the probability of an event at time t given survival until time t. The SF is determined through the CHF as $S(t|\mathbf{x}) = \exp\left(-H(t|\mathbf{x})\right)$.

According to the Cox proportional hazards model [6], the CHF at time t given predictor values $\mathbf{x}$ is defined as

$$H(t|\mathbf{x}, \mathbf{b}) = H_0(t) \exp\left(\mathbf{b}^{\mathrm{T}}\mathbf{x}\right) = H_0(t) \exp\left(\sum\nolimits_{k=1}^{d} b_k x_k\right). \tag{1}$$

Here $H_0(t)$ is the cumulative baseline hazard function; $\mathbf{b}^{\mathrm{T}} = (b_1, ..., b_d)$ is an unknown vector of regression coefficients or parameters. In the framework of the Cox model, the SF $S(t|\mathbf{x}, \mathbf{b})$ is computed as

$$S(t|\mathbf{x}, \mathbf{b}) = \exp(-H_0(t) \exp\left(\mathbf{b}^{\mathrm{T}}\mathbf{x}\right) = (S_0(t))^{\exp\left(\mathbf{b}^{\mathrm{T}}\mathbf{x}\right)}. \tag{2}$$

Here $S_0(t)$ is the baseline SF. It is important to note that functions $H_0(t)$ and $S_0(t)$ do not depend on $\mathbf{x}$ and $\mathbf{b}$.

One of the ways for estimating parameters $\mathbf{b}$ of the Cox model is the Cox partial likelihood function [6].

3   LIME

Let us briefly consider the main ideas behind the LIME method [18]. LIME approximates a black-box model denoted as f with a simple function g in the vicinity of the point of interest $\mathbf{x}$, whose prediction by means of f has to be explained, under condition that the approximation function g belongs to a set of explanation models G, for example, linear models. In order to construct the function g in accordance with LIME, a new dataset consisting of perturbed samples at a local area around the test instance $\mathbf{x}$ is generated, and predictions corresponding to the perturbed samples are obtained by means of the explained model. New samples are assigned by weights $w_{\mathbf{x}}$ in accordance with their proximity to the point $\mathbf{x}$ by using a distance metric, for example, the Euclidean distance or a kernel. The weights are used to enforce locality for the linear model g.

An explanation (local surrogate) model is trained on new generated samples by solving the following optimization problem:

$$\arg\min_{g \in G} L(f, g, w_{\mathbf{x}}) + \varPhi(g). \tag{3}$$

Here L is a loss function, for example, mean squared error, which measures how the explanation is close to the prediction of the black-box model; $\varPhi(g)$ is the model complexity.

As a result, the prediction is explained by analyzing coefficients of the local linear model. The output of LIME, therefore, is a set of important features corresponding to coefficients of the linear model.

4   A General Algorithm of SurvLIME and SurvLIME-Inf

Suppose that there are a training set D and a black-box model which produces an output in the form of the CHF $H(t|\mathbf{x})$ for every new instance $\mathbf{x}$. An idea behind SurvLIME is to approximate the output of the black-box model with the CHF produced by the Cox model for the same input instance $\mathbf{x}$. With this approximation, we get the parameters $\mathbf{b}$ of the approximating Cox model, whose values can be regarded as quantitative impacts on the prediction $H(t|\mathbf{x})$. The largest coefficients indicate the corresponding important features.

Denote the Cox CHF as $H_{\mathrm{Cox}}(t|\mathbf{x}, \mathbf{b})$. Then we have to find such parameters $\mathbf{b}$ that the distance between $H(t|\mathbf{x})$ and $H_{\mathrm{Cox}}(t|\mathbf{x}, \mathbf{b})$ for the considered instance $\mathbf{x}$ would be as small as possible. In order to avoid incorrect results, a lot of nearest points $\mathbf{x}_k$ in a local area around $\mathbf{x}$ is generated. For every $\mathbf{x}_k$, the CHF $H(t|\mathbf{x}_k)$ of the black-box model is obtained as a prediction of the black-box model. Now optimal values of $\mathbf{b}$ can be computed by minimizing the weighted average distance between every pair of CHFs $H(t|\mathbf{x}_k)$ and $H_{\mathrm{Cox}}(t|\mathbf{x}_k, \mathbf{b})$ over all points $\mathbf{x}_k$. Weight w_k assigned to the k-th distance depends on the distance between $\mathbf{x}_k$ and $\mathbf{x}$. Smaller distances between $\mathbf{x}_k$ and $\mathbf{x}$ produce larger weights of distances between CHFs.

It is important to point out that the optimization problem for computing parameters $\mathbf{b}$ depends on the used distance metric between CHFs $H(t|\mathbf{x}_k)$ and

$H_{\text{Cox}}(t|\mathbf{x}_k, \mathbf{b})$. SurvLIME uses the L_2-norm which leads to a convex optimization problem. SurvLIME-Inf uses the L_∞-norm. We will show that this distance metric leads to the linear programming problem whose solution is very simple.

5   Optimization Problem for Computing Parameters

Let $t_0 < t_1 < ... < t_m$ be the distinct times to event of interest, for example, times to deaths from the set $\{T_1, ..., T_n\}$, where $t_0 = \min_{k=1,...,n} T_k$ and $t_m = \max_{k=1,...,n} T_k$. The black-box model maps the feature vectors $\mathbf{x} \in \mathbb{R}^d$ into piecewise constant CHFs $H(t|\mathbf{x})$ such that $H(t|\mathbf{x}) \geq 0$ for all t, $\max_t H(t|\mathbf{x}) < \infty$. Let us introduce the time $T \geq t_m$ in order to restrict $H(t|\mathbf{x})$ and denote $\Omega = [0, T]$.

Interval Ω can be divided into $m + 1$ non-intersecting subsets $\Omega_0, ..., \Omega_m$ such that $\Omega_j = [t_j, t_{j+1})$, $j = 0, ..., m - 1$, $\Omega_m = [t_m, T]$. After introducing the indicator functions $I_j(t)$, which takes value 1 when $t \in \Omega_j$, and 0 otherwise, $H(t|\mathbf{x})$ can be written as follows:

$$H(t|\mathbf{x}) = \sum_{j=0}^{m} H_j(\mathbf{x}) \cdot I_j(t). \tag{4}$$

Here $H_j(\mathbf{x})$ is a part of $H(t|\mathbf{x})$ in interval Ω_j, which does not depend on t because it is constant in this interval. The same can be written for the Cox CHF:

$$H_{\text{Cox}}(t|\mathbf{x}, \mathbf{b}) = H_0(t) \exp\left(\mathbf{b}^{\text{T}}\mathbf{x}\right) = \sum_{j=0}^{m} \left[H_{0j} \exp\left(\mathbf{b}^{\text{T}}\mathbf{x}\right)\right] \cdot I_j(t). \tag{5}$$

It should be noted that the use CHFs for computing the distance between them leads to a complex optimization problem which may be non-convex. Therefore, we proposed to take logarithms of $H(t|\mathbf{x})$ and $H_{\text{Cox}}(t|\mathbf{x}, \mathbf{b})$ denoted as $\phi(t|\mathbf{x})$ and $\phi_{\text{Cox}}(t|\mathbf{x}, \mathbf{b})$, respectively. Since the logarithm is a monotone function, then there hold

$$\phi(t|\mathbf{x}) = \ln H(t|\mathbf{x}) = \sum_{j=0}^{m} (\ln H_j(\mathbf{x})) I_j(t), \tag{6}$$

$$\phi_{\text{Cox}}(t|\mathbf{x}, \mathbf{b}) = \ln H_{\text{Cox}}(t|\mathbf{x}, \mathbf{b}) = \sum_{j=0}^{m} \left(\ln H_j(\mathbf{x}) - \ln H_{0j} - \mathbf{b}^{\text{T}}\mathbf{x}\right) I_j(t). \tag{7}$$

The distance between $\phi(t|\mathbf{x}_k)$ and $\phi_{\text{Cox}}(t|\mathbf{x}_k, \mathbf{b})$ based on the L_∞-norm for every generated point $\mathbf{x}_k$:

$$D_{\infty,k}(\phi, \phi_{\text{Cox}}) = \|\phi(t|\mathbf{x}_k) - \phi_{\text{Cox}}(t|\mathbf{x}_k, \mathbf{b})\|_\infty$$
$$= \max_{t \in \Omega} |\phi(t|\mathbf{x}_k) - \phi_{\text{Cox}}(t|\mathbf{x}_k, \mathbf{b})|. \tag{8}$$

In order to find optimal values of $\mathbf{b}$, the weighted average distance between $\phi(t|\mathbf{x}_k)$ and $\phi_{\text{Cox}}(t|\mathbf{x}_k, \mathbf{b})$ for N generated points $\mathbf{x}_k$ has to be minimized over $\mathbf{b}$, i.e., we have to solve the problem:

$$\min_{\mathbf{b}} \left(\sum_{k=1}^{N} w_k \cdot \max_{t \in \Omega} |\phi(t|\mathbf{x}_k) - \phi_{\text{Cox}}(t|\mathbf{x}, \mathbf{b})|\right). \tag{9}$$

Let us introduce the optimization variables

$$z_k = \max_{t \in \Omega} |\phi(t|\mathbf{x}_k) - \phi_{\text{Cox}}(t|\mathbf{x}_k, \mathbf{b})| . \tag{10}$$

They can be represented as two constraints

$$z_k \geq \phi(t|\mathbf{x}_k) - \phi_{\text{Cox}}(t|\mathbf{x}_k, \mathbf{b}), \ \forall t \in \Omega, \tag{11}$$

$$z_k \geq \phi_{\text{Cox}}(t|\mathbf{x}_k, \mathbf{b}) - \phi(t|\mathbf{x}_k), \ \forall t \in \Omega. \tag{12}$$

Substituting (6)–(7) into (9), using the property that $H_j(\mathbf{x})$ and H_{0j} do not depend on t, and taking into account (11)–(12), we get

$$\min_{\mathbf{b}} \sum_{k=1}^{N} w_k z_k, \tag{13}$$

subject to $\forall t \in \Omega$ and $k = 1, ..., N$,

$$z_k \geq \ln H_j(\mathbf{x}_k) - \ln H_{0j} - \mathbf{b}^{\text{T}} \mathbf{x}_k, \ j = 0, ..., m, \tag{14}$$

$$z_k \geq \mathbf{b}^{\text{T}} \mathbf{x}_k + \ln H_{0j} - \ln H_j(\mathbf{x}_k), \ j = 0, ..., m. \tag{15}$$

Note that term $\mathbf{b}^{\text{T}} \mathbf{x}_k$ does not depend on j. This implies that the constraints can be reduced to the following simple constraints:

$$z_k \geq Q_k - \mathbf{x}_k \mathbf{b}^{\text{T}}, \ k = 1, ..., N, \tag{16}$$

$$z_k \geq \mathbf{x}_k \mathbf{b}^{\text{T}} - R_k, \ k = 1, ..., N. \tag{17}$$

where

$$Q_k = \max_{j=0,...,m} \left(\ln H_j(\mathbf{x}_k) - \ln H_{0j} \right), \tag{18}$$

$$R_k = \min_{j=0,...,m} \left(\ln H_j(\mathbf{x}_k) - \ln H_{0j} \right). \tag{19}$$

Finally, we get the linear optimization problem with objective function (13) and constraints (16)–(17). It has $d + N$ optimization variables ($z_1, ..., z_N$ and $\mathbf{b}$) and $2N$ constraints.

Finally, we write the following scheme of Algorithm 1.

6 Numerical Experiments

6.1 Synthetic Data

In order to get synthetic data, random survival times to events are generated by using the Cox model estimates. For experiments, 1000 covariate vectors $\mathbf{x} \in \mathbb{R}^d$ are randomly generated from the uniform distribution in the d-sphere ($d = 5$) with predefined radius $R = 8$ and center $p = (0, 0, 0, 0, 0)$ of the sphere. There are several methods for the uniform sampling of points $\mathbf{x}$ in the d-sphere, for example, [9]. Bender et al. [3] proposed a method for generating random survival

---

Algorithm 1. The algorithm for computing vector **b** for point **x** in SurvLIME-Inf

---

Require: Training set D; point of interest **x**; the number of generated points N; the black-box survival model for explaining $f(\mathbf{x})$
Ensure: Vector **b** of important features
1: Compute the baseline CHF $H_0(t)$ of the approximating Cox model on dataset D by using the Nelson–Aalen estimator
2: Generate $N - 1$ random nearest points $\mathbf{x}_k$ in a local area around **x**, point **x** is the N-th point
3: Get the prediction of $H(t|\mathbf{x}_k)$ by using the black-box survival model (the function f)
4: Compute weights $w_k = K(\mathbf{x}, \mathbf{x}_k)$ of perturbed points, $k = 1, ..., N$
5: Find vector **b** by solving the convex optimization problem (13), (16)-(17)

---

time data for the Cox model. The Weibull distribution with the scale $\lambda = 10^{-5}$ and shape $v = 2$ parameters is used to generate appropriate survival times because this distribution shares the assumption of proportional hazards with the Cox regression model [3]. Then the following expression can be used for generating survival times [3]:

$$T = \left(\frac{-\ln(U)}{\lambda \exp(\mathbf{b}^{\mathrm{T}} \mathbf{x})} \right)^{1/v}, \tag{20}$$

where U is the random variable uniformly distributed in interval $[0, 1]$.

We take the vector $\mathbf{b}^{\mathrm{T}} = (-0.25, 10^{-6}, -0.1, 0.35, 10^{-6})$ for substituting it into (20). It can be seen that vector **b** has two almost zero-valued elements and three "large" elements which will correspond to important features. Generated values T_i are restricted by the condition: if $T_i > 2000$, then T_i is replaced with value 2000. The event indicator δ_i is generated from the binomial distribution with probabilities $\Pr\{\delta_i = 1\} = 0.9$, $\Pr\{\delta_i = 0\} = 0.1$.

Perturbations can be viewed as a step of the algorithm. According to it, N nearest points $\mathbf{x}_k$ are generated in a local area around **x**. These points are uniformly generated in the d-sphere with some predefined radius $r = 0.5$ and with the center at point **x**. Weights to every point are assigned as follows:

$$w_k = 1 - \left(r^{-1} \cdot \|\mathbf{x} - \mathbf{x}_k\|_2 \right)^{1/2}. \tag{21}$$

To compare vectors **b**, we introduce the following notation: $\mathbf{b}^{\mathrm{model}}$ are coefficients of the Cox model which is used as the black-box model; $\mathbf{b}^{\mathrm{true}}$ are coefficients used for training data generation (see (20)); $\mathbf{b}^{\mathrm{expl}}$ are explaining coefficients obtained by using the proposed explanation algorithm.

One of the aims of numerical experiments is to consider the method behavior by assuming that the black-box model is the Cox model. With these experiments, we have an opportunity to compare the vector $\mathbf{b}^{\mathrm{true}}$ with vectors $\mathbf{b}^{\mathrm{model}}$ and $\mathbf{b}^{\mathrm{expl}}$ because the black-box Cox model as well as the explanation Cox model are expected to have close vectors **b**. We cannot perform the same comparison by using the RSF as a black-box model. Therefore, the results with the RSF will

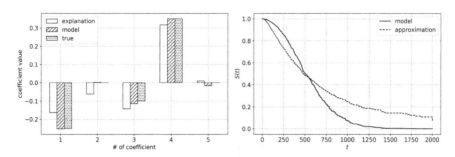

Fig. 1. Results of experiments on synthetic with approximating the black-box Cox model

be compared by considering the proximity of SFs obtained from the RSF and the explanation Cox model.

To evaluate the algorithm, 900 instances are randomly selected from 1000 generated examples for training, and 100 instances are taken for testing. In the test phase, the optimal explanation vector $\mathbf{b}^{\mathrm{expl}}$ is computed for every point from the testing set. In accordance with the obtained vectors $\mathbf{b}^{\mathrm{expl}}$, we depict the mean approximation on the basis of the Euclidean distance between vectors $\mathbf{b}^{\mathrm{expl}}$ and $\mathbf{b}^{\mathrm{model}}$ (for the Cox model) and with Euclidean distance between $H(t_j|\mathbf{x}_i)$ and $H_{\mathrm{Cox}}\left(t_j|\mathbf{x}_i, \mathbf{b}_i^{\mathrm{expl}}\right)$ (for the RSF). In order to get these approximations, a point with the mean approximation is selected among all testing points.

The results of the experiment for the black-box Cox model are depicted in Fig. 1. The left picture shows values of important features $\mathbf{b}^{\mathrm{expl}}$, $\mathbf{b}^{\mathrm{model}}$ and $\mathbf{b}^{\mathrm{true}}$. It can be seen from the picture that the experiment shows the clear coincidence of important features for the model. Right picture in Fig. 1 shows SFs computed by using the black-box Cox model and the Cox approximation.

Similar results for the black-box RSF model are shown in Fig. 2. The important features are not shown in Fig. 2 because RSF does not provide the important features like the Cox model. However, it follows from the SFs in Fig. 2 that the proposed method provides the perfect approximation of the RSF output by the Cox model.

Measures $RMSE_{\mathrm{model}}$ and $RMSE_{\mathrm{true}}$ as functions of the sample size n for SurvLIME and SurvLIME-Inf are provided in Table 1 for comparison purposes. They are defined for the Cox model from n_{test} testing results as follows:

$$RMSE_{\mathrm{type}} = \left(\frac{1}{n_{\mathrm{test}}} \sum_{i=1}^{n_{\mathrm{test}}} \left\| \mathbf{b}_i^{\mathrm{type}} - \mathbf{b}_i^{\mathrm{expl}} \right\|_2 \right)^{1/2}, \tag{22}$$

where is "model" and "true" is substituted into the above expression in place of "type".

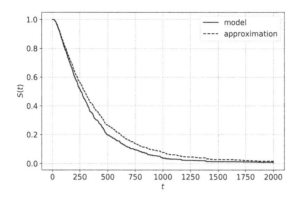

Fig. 2. Results of experiments on synthetic with approximating the black-box Cox model

Table 1. Approximation measures for four cases of using the black-box Cox model by the small amount of data for SurvLIME and SurvLIME-Inf

n	SurvLIME		SurvLIME-Inf	
	$RMSE_{model}$	$RMSE_{true}$	$RMSE_{model}$	$RMSE_{true}$
10	0.719	0.809	0.290	0.575
20	0.659	0.664	0.358	0.460
30	0.347	0.428	0.398	0.432
40	0.324	0.344	0.388	0.451

$RMSE_{model}$ characterizes how the obtained important features coincide with the corresponding features obtained by using the Cox model as the black-box model. $RMSE_{true}$ considers how the obtained important features coincide with the features used for generating the random times to events.

It can be seen from Table 1 that SurvLIME-Inf outperforms SurvLIME for small n, namely, for $n = 10$ and 20. At the same time, this outperformance disappears with increasing n, i.e., when $n = 30$ and 40. This is a very interesting observation which tells us that SurvLIME-Inf should be used when the training set is very small.

The same experiments are carried out for the RSF. They are shown in Fig. 3, where the left column of pictures depicts SFs obtained by means of SurvLIME, the right column of pictures corresponds to SurvLIME-Inf. The k-th row of pictures in Fig. 3 shows results when $n = 10 \cdot k$.

6.2   Real Data

Let us apply the following well-known real datasets to study the method.

The Veterans' Administration Lung Cancer Study (Veteran) Dataset contains data on 137 males with advanced inoperable lung cancer. The subjects

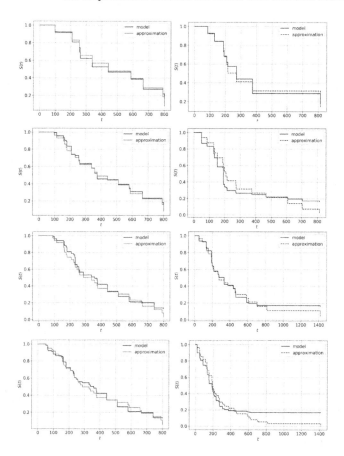

Fig. 3. Comparison of the SFs relationship for SurvLIME (left pictures) and SurvLIME-Inf (right pictures) with using the RSF by 10, 20, 30, 40 training examples

were randomly assigned to either a standard chemotherapy treatment or a test chemotherapy treatment. Several additional variables were also measured on the subjects. The number of features is 6, but it is extended till 9 due to categorical features.

The NCCTG Lung Cancer (LUNG) Dataset records the survival of patients with advanced lung cancer, together with assessments of the patients performance status measured either by the physician and by the patients themselves. The data set contains 228 patients, including 63 patients that are right censored (patients that left the study before their death). The number of features is 8, but it is extended till 11 due to categorical features.

The above datasets can be downloaded via the "survival" R package.

Figure 4 illustrates numerical results for the Veteran dataset. We provide only the case of the mean approximation in order to reduce the number of similar

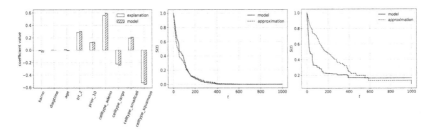

Fig. 4. The mean approximation for the Cox model (the first and the second picture) and the RSF (the third picture) trained on the Veteran dataset

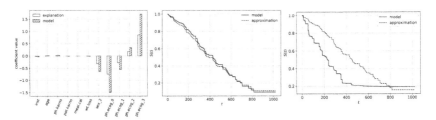

Fig. 5. The mean approximation for the Cox model (the first and the second picture) and the RSF (the third picture) trained on the LUNG dataset

pictures. Figure 4 contains three pictures: the first one illustrates the explanation important features and important features computed by using the Cox model; the second picture shows two SFs for the Cox model; the third picture shows two SFs for the RSF. It follows from Fig. 4 that the method provides appropriate results for the real dataset.

Similar numerical results for the LUNG dataset are shown in Fig. 5.

7 Conclusion

A new modification of SurvLIME using the L_∞-norm for computing the distance between CHFs instead of the L_2-norm has been presented in the paper. The basic idea behind both the methods is to approximate a survival machine learning model at a point by the Cox proportional hazards model which assumes a linear combination of the instance covariates. However, this idea is differently implemented in SurvLIME and SurvLIME-Inf. SurvLIME-Inf extends the set of explanation methods dealing with censored data in the framework of survival analysis. Numerical experiments with synthetic and real datasets have clearly illustrated accuracy and correctness of SurvLIME-Inf.

The main advantage of SurvLIME-Inf is that it uses the linear programming for computing coefficients of the approximating Cox model. This peculiarity allows us to develop new methods taking into account inaccuracy of training data, possible imprecision of data. This is an interesting and important direction for further research. Another problem, which can be solved by using SurvLIME-Inf, is to explain machine learning survival models by using the Cox model with time-dependent covariates. This is also an important direction for further research.

References

1. Ahern, I., Noack, A., Guzman-Nateras, L., Dou, D., Li, B., Huan, J.: Norm-Lime: a new feature importance metric for explaining deep neural networks. arXiv:1909.04200, September 2019
2. Arrieta, A.B., et al.: Explainable artificial intelligence (XAI): concepts, taxonomies, opportunities and challenges toward responsible AI. arXiv:1910.10045, October 2019
3. Bender, R., Augustin, T., Blettner, M.: Generating survival times to simulate cox proportional hazards models. Stat. Med. **24**(11), 1713–1723 (2005)
4. Bou-Hamad, I., Larocque, D., Ben-Ameur, H.: A review of survival trees. Stat. Surv. **5**, 44–71 (2011)
5. Carvalho, D.V., Pereira, E.M., Cardoso, J.S.: Machine learning interpretability: a survey on methods and metrics. Electronics **8**(832), 1–34 (2019)
6. Cox, D.R.: Regression models and life-tables. J. R. Stat. Soc. Ser. B (Methodol.) **34**(2), 187–220 (1972)
7. Fong, R., Vedaldi, A.: Explanations for attributing deep neural network predictions. In: Samek, W., Montavon, G., Vedaldi, A., Hansen, L.K., Müller, K.-R. (eds.) Explainable AI: Interpreting, Explaining and Visualizing Deep Learning. LNCS (LNAI), vol. 11700, pp. 149–167. Springer, Cham (2019). https://doi.org/10.1007/978-3-030-28954-6_8
8. Guidotti, R., Monreale, A., Ruggieri, S., Turini, F., Giannotti, F., Pedreschi, D.: A survey of methods for explaining black box models. ACM Comput. Surv. **51**(5), 93 (2019)
9. Harman, R., Lacko, V.: On decompositional algorithms for uniform sampling from n-spheres and n-balls. J. Multivar. Anal. **101**, 2297–2304 (2010)
10. Kleinbaum, D.G., Klein, M.: Survival Analysis, 3rd edn. Springer, New York (2010)
11. Kovalev, M.S., Utkin, L.V., Kasimov, E.M.: SurvLIME: a method for explaining machine learning survival models. Knowl.-Based Syst. **203**, 106164 (2020)
12. Lee, C., Zame, W.R., Yoon, J., van der Schaar, M.: DeepHit: a deep learning approach to survival analysis with competing risks. In: 32nd Association for the Advancement of Artificial Intelligence (AAAI) Conference, pp. 1–8 (2018)
13. Lundberg, S.M., Lee, S.-I.: A unified approach to interpreting model predictions. In: Advances in Neural Information Processing Systems, pp. 4765–4774 (2017)
14. Mohseni, S., Zarei, N., Ragan, E.D.: A survey of evaluation methods and measures for interpretable machine learning. arXiv:1811.11839, December 2018
15. Molnar, C.: Interpretable machine learning: a guide for making black box models explainable (2019). https://christophm.github.io/interpretable-ml-book/
16. Murdoch, W.J., Singh, C., Kumbier, K., Abbasi-Asl, R., Yua, B.: Interpretable machine learning: definitions, methods, and applications. arXiv:1901.04592, January 2019
17. Nezhad, M.Z., Sadati, N., Yang, K., Zhu, D.: A deep active survival analysis approach for precision treatment recommendations: application of prostate cancer. arXiv:1804.03280v1, April 2018

18. Ribeiro, M.T., Singh, S., Guestrin, C.: Why should I trust You? Explaining the predictions of any classifier. arXiv:1602.04938v3, August 2016
19. Ribeiro, M.T., Singh, S., Guestrin, C.: Anchors: high-precision model-agnostic explanations. In: AAAI Conference on Artificial Intelligence, pp. 1527–1535 (2018)
20. Rudin, C.: Stop explaining black box machine learning models for high stakes decisions and use interpretable models instead. Nat. Mach. Intell. **1**, 206–215 (2019)
21. Wachter, S., Mittelstadt, B., Russell, C.: Counterfactual explanations without opening the black box: automated decisions and the GPDR. Harvard J. Law Technol. **31**, 841–887 (2017)
22. Wang, P., Li, Y., Reddy, C.K.: Machine learning for survival analysis: a survey. arXiv:1708.04649, August 2017
23. Zafar, M.R., Khan, N.M.: DLIME: a deterministic local interpretable model-agnostic explanations approach for computer-aided diagnosis systems. arXiv:1906.10263, June 2019

Unsupervised Neural Aspect Extraction with Related Terms

Timur Sokhin$^{(\boxtimes)}$, Maria Khodorchenko , and Nikolay Butakov

ITMO University, St. Petersburg, Russia
qwinpin@gmail.com, mkhodorchenko@niuitmo.ru, alipoov.nb@gmail.com

Abstract. The tasks of aspect identification and term extraction remain challenging in natural language processing. While supervised methods achieve high scores, it is hard to use them in real-world applications due to the lack of labelled datasets. Unsupervised approaches outperform these methods on several tasks, but it is still a challenge to extract both an aspect and a corresponding term, particularly in the multi-aspect setting. In this work, we present a novel unsupervised neural network with convolutional multi-attention mechanism, that allows extracting pairs (aspect, term) simultaneously, and demonstrate the effectiveness on the real-world dataset. We apply a special loss aimed to improve the quality of multi-aspect extraction. The experimental study demonstrates, what with this loss we increase the precision not only on this joint setting but also on aspect prediction only.

Keywords: Unsupervised training · Neural networks · Aspect extraction · Aspect term extraction

1 Introduction

Unsupervised aspect extraction is an essential part of natural language processing and usually solved using topic modelling approaches, which have proven themselves in this task. In general, aspect extraction aims to identify the category or multiple categories of a given text. Aspect is a certain topic, term, in turn, is a particular word or word combination in a text that refers to a particular topic. Previous unsupervised approaches achieved significant improvement in the task of aspect extraction. The joint task of the aspect and the aspect term pairs extraction is still a challenge for natural language processing. For example: in the sentence "Best Pastrami I ever had and great portion without being ridiculous" the aspect and aspect term pairs "Food: Pastrami" and "Food: portion".

Most of the existing approaches apply two-stage extraction: aspect extraction first and then aspect term extraction based on the known aspect. We propose a conjoint solution based on the convolutional multi-attention mechanism (CMAM). The CMAM was inspired by Inception-block in computer vision [14], where kernels of different sizes allow incorporating features from different levels

© Springer Nature Switzerland AG 2020
A. Filchenkov et al. (Eds.): AINL 2020, CCIS 1292, pp. 75–86, 2020.
https://doi.org/10.1007/978-3-030-59082-6_6

of localisation. The sentence representations built with CMAM capture the features, which are used for aspect predictions, while the attention detects related terms. Also, the convolutional attention does not require much additional time to infer the result, which is vital for the real-world application. In order to increase the quality of multi-aspect extraction, we propose a novel loss function - triplet-like aspect spreading (TLAS), which maximises the distance between top-N aspect-based sentence representations and minimises the distance between these representations and corresponding aspect vectors. This approach allows achieving close to the state-of-the-art results in aspect extraction with the ability to extract their terms. In summary, the contributions of this paper are:

- CMAM; convolutional multi-attention mechanism, which is aimed to build sentence vector representation and to extract aspect terms.
- TLAS; loss function for aspect probabilities distribution modifying.
- The experimental study of the proposed model on SemEval-2016 Restaurant dataset and Citysearch corpus.

2   Related Work

Retrieval of aspects is one of the main tasks in the natural language processing, which can be used as a subtask in the analysis of sentiment, classification of documents, tasks to support decision-making. In 2014, the Aspect Based Sentiment Analysis task was launched [13]. Since then, significant results have been achieved both in extracting the aspect itself and in extracting its associated term [4,5,15]. Supervised-learning methods are effective but difficult to apply to real-world tasks for which there is no available dataset. Modern achievements in language models, such as BERT [7,18], can significantly reduce the amount of data required. On the other hand, some approaches allow solving a task without labelled data - unsupervised learning.

In some works, the authors propose to combine the use of rule-based and supervised methods - automatic creation of a labelled dataset [8,17]. However, in this case, there is a high probability of errors in the dataset itself.

With respect to learning without a teacher, first of all, it is necessary to note topic modelling [1,19]: statistical models which allow forming sets of topics based on word occurrence frequency in documents. Such methods are probabilistic, do not take into account the semantics of the text, the order of words in the sentence. Such models also allow to define the presence of many aspects in the text, but they are not possible to identify which part of the document is responsible for a particular topic.

Different neural models also solve a problem in an unsupervised setting and allow the semantics of text data to be taken into account [9–11,16]. The latter models are designed to provide a set of aspects during the learning process, which are then assigned to the individual documents. The given set of aspects is presented in some vector space [12], and each aspect is interpreted as the words nearest to the given vector. At the expense of the attention mechanism such models also allow to define the term in the text which is directly connected with aspects of the text. The drawback of the previous works is that it is implied that the text contains only one aspect and the term related to it. We propose a model that can take into account the presence of many aspects in the text, each of which corresponds to a different term.

3   The Proposal

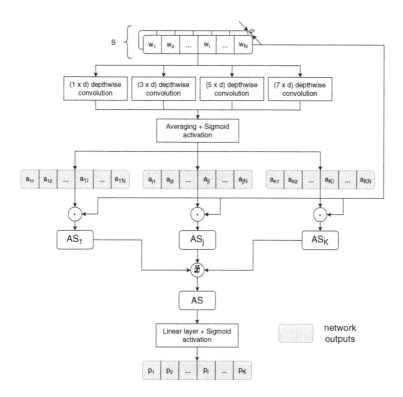

Fig. 1. Schema of the proposed approach to aspect and aspect term extraction.

The general neural attention model for aspect extraction is based on ABAE idea [9], which significantly outperformed previous methods of unsupervised aspect predictions. In this model, the set of aspects is learned in the same embedding space with words and can be easily interpreted by a human. However, despite the similarities at the level of aspect matrix generation, our approach differs at the core of attention mechanism: we aim to capture the most relevant information from the sentence for each aspect encountered independently - Fig. 1. The main idea can be shown as:

1. Build attention for each aspect using CMAM.
2. Build sentence representations for each of these attentions - attentioned sentence representations (AS_j).
3. Build attentioned averaged sentence (AS) representation with AS_j averaging.
4. Infer the weights of aspects.

After we have the weights of aspects, we can select the aspect term(-s) from corresponding attentions.

3.1 Model

Convolutional Multi-attention Mechanism. CMAM makes it possible to extract term for each individual aspect using simple convolutional approach. The set of convolutions with different kernel sizes allows analysing each word of the sentence on different levels to take into account the local and the global context. CMAM consists of F depthwise convolutions $Conv_j$ with K channels, each channel corresponds to its own aspect:

$$A' = \frac{1}{F} \sum_{i=1}^{F} S * Conv_j \tag{1}$$

$$A = \frac{1}{1 + e^{-A'}} \tag{2}$$

$$AS_j = A_j \cdot S \tag{3}$$

$$AS = \frac{1}{K} \sum_{j=1}^{K} AS_j \tag{4}$$

where j means aspect index; $S \in \mathbb{R}^{N \times d}$ - sentence, represented as a set of N word embeddings w_i with dimension d; $A \in \mathbb{R}^{N \times K}$ is an attention matrix with K - number of aspects; AS_j is attentioned sentence representation over a given aspect j; AS - averaged sentence representation.

Each of the individual AS_j is associated with its aspect, which further allows obtaining a different term for each aspect.

Aspect Prediction. Averaging in Eq. 4 is used for the aspect probabilities prediction using simple linear transformation (Eq. 5).

$$p = \frac{1}{1 + e^{-(wAS+b)}} \tag{5}$$

where w - learnable weights, b - learnable bias.

Aspects of the text are not collectively exhaustive events, there may be several at the same time. In order to take into account this feature, we use Sigmoid activation instead of Softmax.

Sentence Reconstruction. Sentence reconstruction is required for the model training in an unsupervised way. Like an auto-encoder, we match the reconstructed sentence RS with the averaged embeddings of the original sentence. The reconstructed sentence, in this case, is the linear combination of AS and aspect embedding matrix $AEM \in \mathbb{R}^{K \times d}$, which gives us a weighted sum of all aspects. AEM is a trainable parameter of the neural network.

Aspect and Aspect Term Inferring. The final output of the neural network: weights for each aspect and weights of attention for each of them. Since it is assumed that a sentence can contain more than one aspect as well as none, we apply the following prediction scheme: first, we select aspects with a weight greater than q-th quantile q_{as} of all weights and then select Top-2 aspects as the final result.

For each selected aspect it is necessary to define the aspect terms from the text: using the weight of the attention layer Eq. 2 corresponding to its aspect, we apply the same approach - we select the words whose weight is greater than the q-th quantile q_{at} of all words in this sentence and then select Top-1. q_{as} and q_{at} are hyperparameters of the model.

3.2 Training Objective

Our approach consist of loss functions:

- Hinge loss H, which demonstrates better result comparing with Triplet loss. We minimise the reconstruction error in the form of the inner product between RS and averaged embeddings of original sentence S and maximise the difference between RS and randomly sampled negative sentences, which are formed as an averaging of words embeddings per sample - in the same way as the training of the Siamese networks [6].
- Orthogonality regularisation aimed to make the aspect embeddings more representative and unique.
- Triplet-like aspect spreading - since we want to be able to detect more than one aspect, TLAS makes top-N aspects more unique, helps to avoid repetitions of similar aspects.

Orthogonality Loss. Orthogonal loss in Eq. 6 allows us to make the vectors of individual aspects more unique. However, our experiments demonstrate that, when orthogonality value is close to zero, aspects can "degenerate" and cover only a small area of a particular category. This effect leads to an increase in the quality of determination of the aspects themselves but reduces the overall efficiency of pair extraction (aspect: aspect term).

$$U = \lambda \left(\left\| AEM \cdot AEM^T - I \right\| - s \right) \tag{6}$$

where s is an offset, which determines to which value U will be optimised, λ is a weight of this loss, I - identity matrix.

TLAS. The idea is that the top-N aspect of each sentence should cover different categories:

$$\sqrt{(AS_j - AS_l)^2} \to \infty \tag{7}$$

where j and l are indexes of the Top-2 aspects. We use Top-2, because of the mean number of aspects equal to 2.25 among all dataset with more than 1 aspect labelled by a human reviewer.

From the other hand, each Top-2 AS_j must be closer to corresponding aspect vector from aspect embedding matrix - AEM_j. The overall TLAS loss is formulated as:

$$T = max\left(0, 1 + \sqrt{\sum_{k=1}^{d}(AS_{jk} - AEM_{jk})^2} - \sqrt{\sum_{k=1}^{d}(AS_{jk} - AS_{lk})^2}\right) \tag{8}$$

Summary loss function L is:

$$L = H + U + T \tag{9}$$

4   Experiments

4.1   Datasets

Out experiments conducted as unconstrained SemEval task: we use the whole Citysearch corpus to train our model and to evaluate aspect extraction efficiency and use SemEval-2016 Restaurant reviews dataset [13] to evaluate the aspect and aspect term co-extraction efficiency. The joint aspect and aspect term extraction was a part of task 5 "Aspect Based Sentiment Analysis". Due to under-representation or ambiguity of some categories, the final evaluation labels presented in the Table 1.

Table 1. Map of origin labels and the new. Categories, which are not used for evaluation: Prices, Location, Restaurant, Misc

New label	Origin label	Size
Ambience	Ambience general	82
Food	Food Quality	447
	Food Style options	
	Drinks Quality	
	Drinks Style options	
Staff	Service general	157
Multi-labels	All samples with	337
	More than one label	

Restaurant category was omitted because of the ambiguity of terms selected by SemEval reviewers - 45.8% of all terms for this aspect are "place" and "restaurant". All samples in these categories contain one pair (aspect: aspect term). Samples with more than one pair are grouped in the multi-label category.

4.2 Experimental Settings

We use word2vec model trained on Citysearch reviews with embedding size 200, window size 10 and negative size 20 for embedding matrix initialisation. Texts filtered from stop words, punctuation. Total vocabulary size is 9000 that covers 77.1% of SemEval dataset. Words out of vocabulary marked as "unknown", and if they were predicted, we consider them as an error. While in works related to unsupervised aspect extraction [3,9,20] the optimal number of aspects is 14 for the Restaurant domain, our experiments demonstrate that the best result is achieved using 30 pre-defined aspects initialised with centroids of k-means clustering over embedding matrix. Model is trained for 5 epochs using Adam optimiser: learning rate 0.0005, default betas 0.9 and 0.999; with batch size 64. Orthogonal loss weight is set to 0.5, offset 0.3. q_{as} and q_{at} are 0.1.

4.3 Evaluation Settings

Table 2. Inferred aspects, their representative words and gold-labels mapping. Price and misc. gold-labels were omitted due to ambiguity of their meanings.

Generated aspect	Representative words	Gold-label
Drinks	cocktail, liquor, beer	Food
Ingredients	ceviche, oxtail, saffron	
Misc. Food	biscuit, onion, bun	
Menu	menu, selection, option	
Cuisine style	food, cuisine, fusion	
Geography	mexican chinese japanese	
Staff	manager, maitre, politely	Staff
Staff	service, waitstaff, staff	
Attitude	attentive, friendly, polite	
Environment	couch, fireplace, patio	Ambience
Style	retro, deco, overhead	
Price	overpriced, average	Price
Location	manhattan, brooklyn, ny	Misc
Opinion terms	great, fantastic, amazing	Misc
Place	place restaurant spot	Restaurant

The nearest words can describe the generated aspect embeddings in embeddings space. For the task of aspect extraction, each of 30 generated aspects was mapped with one of the six gold-standard label - Table 2.

For aspect and aspect term extraction the partial coincidence of the predicted term and gold-term is also considered as the true positive if the aspect is correct. That is also done due to the ambiguity of data labels.

4.4   Aspect Extraction Results

Topic (aspect) modelling approaches such as LDA-based [2] retain their significance in this task, and they can still compete with modern methods. The results of BTM [19], SERBM [16] are taken from [9]. Approaches based on neural networks are also actively developing: ABAE, AE-CSA. Their results are also taken from the authors' publications [9, 11].

Compared to the others, our approach demonstrates close to state-of-the-art results - Table 3. Also, we analyse the result of our model without TLAS loss. Despite a rather high efficiency, the decrease is noticeable both in precision and recall. In addition, $\lambda = 0.0$ in Eq. 6 leads to even greater reduction of F1-score despite high precision value. Further, we demonstrate that in the joint problem, that leads to more significant deterioration.

Table 3. Aspect extraction results.

Aspect	Model	Precision	Recall	F1
Food	BTM	0.933	0.745	0.816
	SERBM	0.891	0.854	0.872
	ABAE	0.953	0.741	0.828
	AE-CSA	0.903	0.926	0.914
	Ours	0.887	0.945	0.915
	Ours withoutTLAS	0.910	0.890	0.900
	Ours, $\lambda = 0.0$	0.817	0.909	0.860
Staff	BTM	0.828	0.579	0.677
	SERBM	0.819	0.582	0.680
	ABAE	0.802	0.728	0.757
	AE-CSA	0.804	0.756	0.779
	Ours	0.804	0.676	0.735
	Ours without TLAS	0.668	0.662	0.665
	Ours, $\lambda = 0.0$	0.852	0.554	0.671
Ambience	BTM	0.813	0.599	0.685
	SERBM	0.805	0.592	0.682
	ABAE	0.815	0.698	0.740
	AE-CSA	0.768	0.773	0.770
	Ours	0.763	0.757	0.760
	Ours without TLAS	0.672	0.733	0.701
	Ours, $\lambda = 0.0$	0.817	0.677	0.741

4.5   Aspect and Aspect Term Extraction Results

We provide the experimental results for different aspects in Table 4. We could not find a strong dependency between the quality of aspect detection and term extraction. We can see the influence of TLAS loss and orthogonality shift, especially at the multi-label subtask. The drawback of our approach is the weight sharpness of the term: non-aspect terms weights are almost zero. Solving this problem will allow us to more accurately handle the search for long phrases.

Table 4. F1 score for aspect and aspect term extraction for different model variations.

Aspect	Ours (F1)	Without TLAS (F1)	$\lambda = 0.0$ (F1)
Ambience	0.279	0.157	0.227
Food	0.426	0.196	0.349
Staff	0.501	0.165	0.406
Multi-label	0.296	0.103	0.173
Micro-average	0.369	0.149	0.272

Also, we must note the difficulty of evaluating of unsupervised approaches on SemEval-2016 restaurant dataset. Supervised approaches work well with strict rules and control (labels); unsupervised is not limited by outside and can give result, which is not obvious for a human reviewer at first sight.

5   Conclusions

In this paper, we presented a new unsupervised neural model with the convolutional multi-attention mechanism for aspect search with related term extraction. In the experimental study, we show the efficiency of our model in the task of aspect extraction compared to the other state-of-the-art approaches and its possibility of correct aspect and aspect term joint identification. In Fig. 2 we provide two examples of sentences and their aspect and aspect term.

Prediction	Gold-label	Rude	service	medicore	food	there	are	tons	of	restaurants	NY	stay	away
Location	-	0,001	0,001	0,001	0	0	0	0	0	0	0,061	0	0,002
Staff	Staff	0,013	0,297	0	0	0	0	0	0	0	0	0,002	0,002

Prediction	Gold-label	The	wine	list	was	extensive	-	though	the	staff	did	not	seem	knowledgeable
Staff	Staff	0,031	0,017	0,001	0	0	0	0,001	0	0,495	0	0	0	0,002
Drinks	Drinks	0	0,603	0,022	0	0,016	0	0,003	0	0,007	0	0	0	0,002

Fig. 2. Examples of aspect and attention extraction.

The further work can be directed to the enhancement of the inferring procedure to incorporate multiple aspects with high predicted probability within the same category and their terms and also to modification of attention mechanism to achieve more smooth weights distribution between words of the sentence. It is also possible to replace the current method of vector representation generation with more advanced methods.

We also want to note that the evaluation of the quality of such algorithms strongly depends on the labeling used in test datasets: on the rules by which the labeling is made, on subjective errors in its preparation. In the process of experiments with the method of extraction of (aspect: aspect term) pairs proposed by us, we noticed that some predictions which are incorrect or partially correct from the point of view of labeling, are an adequate solution for these examples. This explains our decision to use in the evaluation of metrics in Sect. 4.5 not only the exact coincidence of the term, but also the partial intersection. Further work in this direction may include both expanding the existing data sets and changing the approach to labeling such types of data.

Acknowledgements. This research is financially supported by The Russian Science Foundation, Agreement #20-11-20270.

References

1. Blei, D.M., Ng, A.Y., Jordan, M.I.: Latent dirichlet allocation. In: Dietterich, T.G., Becker, S., Ghahramani, Z. (eds.) Advances in Neural Information Processing Systems 14, pp. 601–608. MIT Press (2002)
2. Blei, D.M., Ng, A.Y., Jordan, M.I.: Latent dirichlet allocation. J. Mach. Learn. Res. **3**, 993–1022 (2003)
3. Brody, S., Elhadad, N.: An unsupervised aspect-sentiment model for online reviews. In: NAACL HLT (2010)
4. Brun, C., Perez, J., Roux, C.: XRCE at SemEval-2016 task 5: feedbacked ensemble modeling on syntactico-semantic knowledge for aspect based sentiment analysis. In: Proceedings of the 10th International Workshop on Semantic Evaluation (SemEval-2016), pp. 277–281. Association for Computational Linguistics, San Diego, California (2016). https://doi.org/10.18653/v1/S16-1044
5. Çetin, F.S., Yıldırım, E., Özbey, C., Eryiğit, G.: TGB at SemEval-2016 task 5: multi-lingual constraint system for aspect based sentiment analysis. In: Proceedings of the 10th International Workshop on Semantic Evaluation (SemEval-2016), pp. 337–341. Association for Computational Linguistics, San Diego, California (2016). https://doi.org/10.18653/v1/S16-1054
6. Chopra, S., Hadsell, R., LeCun, Y.: Learning a similarity metric discriminatively, with application to face verification. In: 2005 IEEE Computer Society Conference on Computer Vision and Pattern Recognition (CVPR 2005). vol. 1, pp. 539–546 (2005)
7. Devlin, J., Chang, M.W., Lee, K., Toutanova, K.: Bert: pre-training of deep bidirectional transformers for language understanding. In: Proceedings of the 2019 Conference of the North American Chapter of the Association for Computational Linguistics: Human Language Technologies, Volume 1 (Long and Short Papers), pp. 4171–4186 (2019)

8. Giannakopoulos, A., Musat, C., Hossmann, A., Baeriswyl, M.: Unsupervised aspect term extraction with b-LSTM & CRF using automatically labelled datasets. In: Proceedings of the 8th Workshop on Computational Approaches to Subjectivity, Sentiment and Social Media Analysis, pp. 180–188. Association for Computational Linguistics, Copenhagen, Denmark, September 2017. https://doi.org/10.18653/v1/W17-5224

9. He, R., Lee, W.S., Ng, H.T., Dahlmeier, D.: An unsupervised neural attention model for aspect extraction. In: Proceedings of the 55th Annual Meeting of the Association for Computational Linguistics (Volume 1: Long Papers), pp. 388–397. Association for Computational Linguistics, Vancouver, Canada (2017). https://doi.org/10.18653/v1/P17-1036

10. He, R., Lee, W.S., Ng, H.T., Dahlmeier, D.: An interactive multi-task learning network for end-to-end aspect-based sentiment analysis (2019)

11. Luo, L., et al.: Unsupervised neural aspect extraction with sememes. In: Proceedings of the Twenty-Eighth International Joint Conference on Artificial Intelligence, IJCAI 2019, pp. 5123–5129. International Joint Conferences on Artificial Intelligence Organization (2019). https://doi.org/10.24963/ijcai.2019/712

12. Mikolov, T., Sutskever, I., Chen, K., Corrado, G.S., Dean, J.: Distributed representations of words and phrases and their compositionality. In: Burges, C.J.C., Bottou, L., Welling, M., Ghahramani, Z., Weinberger, K.Q. (eds.) Advances in Neural Information Processing Systems 26, pp. 3111–3119. Curran Associates, Inc. (2013)

13. Pontiki, M., et al.: SemEval-2016 task 5: aspect based sentiment analysis. In: Proceedings of the 10th International Workshop on Semantic Evaluation (SemEval-2016), pp. 19–30. Association for Computational Linguistics, San Diego, California, June 2016. https://doi.org/10.18653/v1/S16-1002

14. Szegedy, C., Vanhoucke, V., Ioffe, S., Shlens, J., Wojna, Z.: Rethinking the inception architecture for computer vision. In: 2016 IEEE Conference on Computer Vision and Pattern Recognition (CVPR), pp. 2818–2826 (2016)

15. Toh, Z., Su, J.: NLANGP at SemEval-2016 task 5: improving aspect based sentiment analysis using neural network features. In: Proceedings of the 10th International Workshop on Semantic Evaluation (SemEval-2016), pp. 282–288. Association for Computational Linguistics, San Diego, California (2016). https://doi.org/10.18653/v1/S16-1045

16. Wang, L., Liu, K., Cao, Z., Zhao, J., de Melo, G.: Sentiment-aspect extraction based on restricted boltzmann machines. In: Proceedings of the 53rd Annual Meeting of the Association for Computational Linguistics and the 7th International Joint Conference on Natural Language Processing (Volume 1: Long Papers), pp. 616–625. Association for Computational Linguistics, Beijing, China (2015). https://doi.org/10.3115/v1/P15-1060

17. Wu, C., Wu, F., Wu, S., Yuan, Z., Huang, Y.: A hybrid unsupervised method for aspect term and opinion target extraction. Knowl.-Based Syst. 148, 66–73 (2018). https://doi.org/10.1016/j.knosys.2018.01.019

18. Xu, H., Liu, B., Shu, L., Yu, P.S.: Bert post-training for review reading comprehension and aspect-based sentiment analysis. In: Proceedings of the 2019 Conference of the North American Chapter of the Association for Computational Linguistics, June 2019
19. Yan, X., Guo, J., Lan, Y., Cheng, X.: A biterm topic model for short texts. In: Proceedings of the 22nd International Conference on World Wide Web, pp. 1445–1456. WWW 2013, Association for Computing Machinery, New York (2013). https://doi.org/10.1145/2488388.2488514
20. Zhao, X., Jiang, J., Yan, H., Li, X.: Jointly modeling aspects and opinions with a MaxEnt-LDA hybrid. In: Proceedings of the 2010 Conference on Empirical Methods in Natural Language Processing, pp. 56–65. Association for Computational Linguistics, Cambridge, MA (2010)

Predicting Eurovision Song Contest Results Using Sentiment Analysis

Iiro Kumpulainen, Eemil Praks, Tenho Korhonen, Anqi Ni, Ville Rissanen, and Jouko Vankka[(✉)]

Department of Military Technology, National Defence University, Helsinki, Finland
jouko.vankka@mil.fi

Abstract. Over a million tweets were analyzed using various methods in an attempt to predict the results of the Eurovision Song Contest televoting. Different methods of sentiment analysis (English, multilingual polarity lexicons and deep learning) and translating the focus language tweets into English were used to determine the method that produced the best prediction for the contest. Furthermore, we analyzed the effect of sampling tweets during different periods, namely during the performances and/or during the televoting phase of the competition. The quality of the predictions was assessed through correlations between the actual ranks of the televoting and the predicted ranks. The prediction was based on the application of an adjusted Eurovision televoting scoring system to the results of the sentiment analysis of tweets. A predicted rank for each performance resulted in a Spearman ρ correlation coefficients of 0.62 and 0.74 during the televoting period for the lexicon sentiment-based and deep learning approaches, respectively.

Keywords: Natural language processing · Sentiment analysis · Social media text analysis

1 Introduction

The Eurovision Song Contest is an annual musical performance competition for mostly European countries. Each country has an artist who performs for a maximum of three minutes. After the performances, the viewers in the participating countries can vote via call, text message, or app. Over the course of the event, fans post tweets related to the contest. This paper provides an analysis of tweets posted during the grand finals of the Eurovision Song Contest held on May 18, 2019.

The Eurovision Song Contest provides an opportunity to evaluate methods for analyzing the opinions of the general public on social media. We attempted to predict the results of the contest, assess the quality of the predictions, and find patterns of voting preferences among countries by analyzing tweets. In the future, similar social media analysis could be used to predict the results of elections, and analyze public opinion on important topics. In particular, this paper focuses on

© Springer Nature Switzerland AG 2020
A. Filchenkov et al. (Eds.): AINL 2020, CCIS 1292, pp. 87–108, 2020.
https://doi.org/10.1007/978-3-030-59082-6_7

comparing public opinion between target entities, as well as studying differences in views between different regions.

Since there are some words and combinations of words that are used in the paper but the meaning might not be clear, a dictionary is provided in the appendix.

2   Related Work

There are several studies related to modeling the Eurovision votes [2,5,6,12,13, 15,24–26,31,32]. Voting behavior in the Eurovision contest is at least partially driven by cultural geography, host county effect, cultural, linguistic, historical, and political proximities, as well as characteristics of performance, such as song lyrics, music, language, size of the performance group, and linguistic components (e.g., cultural voting blocs and neighborhood voting). Demergis represents the first attempt to predict Eurovision viewer opinions by studying their reactions on social media [7]. The study analyzed tweets using a naive Bayes classifier for sentiment analysis in English and Spanish. The total number of positive tweets for each country was counted to determine a final score for that country. In comparison, we used a simple sentiment lexicon-based approach as well as a deep learning-based sentiment analysis method. Furthermore, we applied additional rules (i.e., the adjusted Eurovision televoting scoring system) after the sentiment analysis, and analyzed the tweets at different time intervals.

3   Method

The analysis was conceptually divided into the following steps:

A. Collection of Eurovision tweets
B. Identification of the source country
C. Tokenization of tweets
D. Identification of target country
E. Sentiment analysis
F. Tallying of final results

A flowchart visualizing the entire process is presented in Fig. 1.

3.1   Collection of Eurovision Tweets

Past tweets are directly accessible through the Twitter API[1]. Eurovision tweets were collected by Demergis [7] by specifying Eurovision-related hashtags. Hashtags of Eurovision tweets include variations on the phrase "Eurovision Song

---

[1] https://developer.twitter.com/en/docs/tweets/post-and-engage/overview   last accessed Jan. 10, 2020.

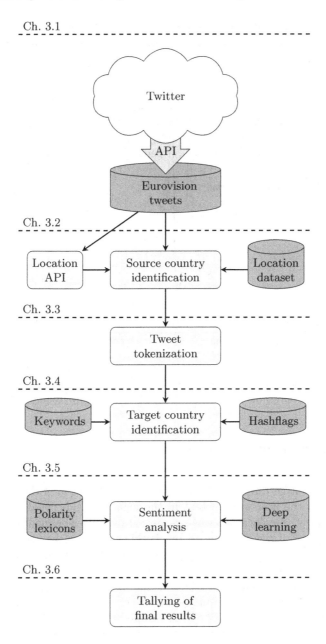

Fig. 1. The analysis procedure

Contest" (#eurovision, #eurovision2019, #esc2019, #esc19, #eurovisionsong-contest) along with the slogan of the 2019 competition, "Dare to Dream" (#dare-todream). The time interval of the tweets was determined to be 19:14 GMT to

22:08 GMT, which includes the performance period (19:14 GMT to 21:10 GMT) and the televoting period (21:10 GMT to 22:08 GMT), but excludes the presentation of the official results (see Fig. 2).

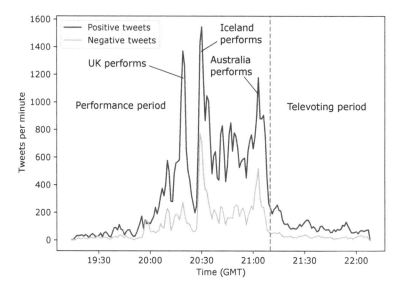

Fig. 2. Positive and negative tweets with identified target and source country per minute during the performance and televoting periods of Eurovision 2019

3.2   Identification of the Source Country

The country from which the tweet was sent (referred to as the source country) was determined either directly from the location information of the tweet (geolocation) or from the user-written text in the location field, usually in the form of 'City, Country.' Geolocation was prioritized over user-defined location. The user-defined location was determined by checking whether the user-written text contained the name of a participating country or the name of a region or city in any language. To match regions and cities with their respective countries, we used the GeoNames dataset[2] to obtain the names of regions and cities with at least 5,000 inhabitants in countries that were allowed to vote in Eurovision 2019. If different regions or cities had the same name, or the user-written location text included names of areas in different countries, we chose the region with the highest population. Tweets with no identified source country were discarded from the dataset.

---

[2] http://download.geonames.org/export/dump/ last accessed Mar. 25, 2020.

3.3   Tweet Tokenization

The tweet texts were split into tokens by parsing the words, emoticons, and emojis in the text. This process is called *tokenization* and was achieved by applying TweetTokenizer from a Python library nltk 3.4.5[3]. This converted the tweet text into lowercase and separated punctuation from words, keeping the emoticons and hashtags intact. The deep learning model used had its own implementation of tokenization.

3.4   Identification of the Target Country

During the contest broadcast, a list of hashtags comprising three-letter abbreviations for each finalist country was displayed. The viewers were encouraged to use these hashtags, referred to as "hashflags," when tweeting about the performance of a country. For example, the hashflag for Sweden was #SWE. These hashflags are listed in Table 1. The hashflags serve as a way to identify the *target country*, or the country to which a tweet referred.

Table 1. Eurovision 2019 hashflags

Albania #ALB	Germany #GER	Russia #RUS
Australia #AUS	Greece #GRE	San Marino #SMR
Azerbaijan #AZE	Iceland #ISL	Serbia #SRB
Belarus #BLR	Israel #ISR	Slovenia #SLO
Cyprus #CYP	Italy #ITA	Spain #ESP
Czech Republic #CZE	Malta #MLT	Sweden #SWE
Denmark #DEN	Netherlands #NED	Switzerland #SUI
Estonia #EST	North Macedonia #MKD	United Kingdom #UK, #GBR
France #FRA	Norway #NOR	

The target country was identified by checking whether any of the words in the tweet text belonged to a keyword list. There was a keyword list for each finalist country, which comprised hashflags and keywords (abbreviations of the country and performer) in the format shown in Table 2.

---

[3] https://www.nltk.org/api/nltk.tokenize.html last accessed Jan. 10, 2020.

Table 2. Keyword examples for Sweden

Keyword type	Example
Country name hashtag	#Sweden
ISO-2 abbreviation hashtag	#SE
ISO-3 abbreviation hashtag	#SWE
Country name in different languages	Schweden
Country flag emoji	🇸🇪
Country demonym	Swede
Other adjective	Swedish
Performer stage name	John Lundvik

If a word belonged to a keyword list, the tweet was associated with the country of the keyword list. If no target country was found, the tweet was rejected. If more than one country was identified as the result of this keyword scanning process, the tweet was also discarded as more sophisticated analysis methods would be required to find the true target country or analyze the sentiment toward each target country. For examples of such cases, see "Multiple" in Table 3.

3.5 Sentiment Analysis

The goal was to identify whether a tweet was positive or negative toward the target country. We used two different algorithms for this purpose: a simple lexicon-based algorithm, and a state-of-the-art deep learning sentiment analysis method. The lexicon-based algorithm used for sentiment analysis evaluated whether an individual word, emoji, or emoticon in the text belonged to a list of known words, emojis, and emoticons with polarity. For each tweet, the sentiment value of the matching tokens was summed, resulting in either a positive or a negative tweet.

As the deep learning method, we used a state-of-the-art method called RoBERTa (Robustly Optimized BERT Pretraining Approach) [18], which is based on Google's language representation model BERT (Bidirectional Encoder Representations from Transformers) [8]. The base RoBERTa model was pretrained on over 160 GB of English text such as English Wikipedia and news articles [18]. For sentiment analysis, the language model was fine-tuned on Stanford Sentiment Treebank v2 dataset [30], which includes sentiment scores by human annotators for 215,154 phrases generated from movie reviews. The RoBERTa model is available in HuggingFace's Transformers Python library[4].

The information for tweets from the Twitter API includes the identified language of each tweet. Based on this language information, three approaches were used in the lexicon-based sentiment analysis: (a) only English tweets, (b) English tweets + tweets translated from focus languages to English with the Google Cloud Translate API and (c) multilingual polarity lexicons. Each English or

---

[4] https://huggingface.co/textattack/roberta-base-SST-2 last accessed July 28, 2020.

Table 3. Country identification and sentiment analysis examples

Tweet text	Country identified	Lexicon sentiment	RoBERTa sentiment
Ooh Norway, I like this! #NOR #Eurovision	Norway	1	Positive
So far #Eurovision is mostly lacking in humor. #SWE is no exception	Sweden	−1	Positive
"Return to Your Land"? Borderline racist if I ever heard it. #ALB #Eurovision	Albania	−1	Negative
This song is so beautiful, simple and passionate I would love to see it do well #svn #eurovision	Slovenia	4.293	Positive
Of course I'll be rooting for Duncan tonight but the Czech are absolutely my fave! #ESC19	Multiple	1.293	Positive
This is very cross cultural - the Czech guy sings with an Aussie accent? I need to drink some more French wine, quick... #Eurovision	Multiple	0	Negative
Very cool staging, but the song position means it'll be forgotten #Eurovision	None	1.293	Negative
Uninspired, mediocre and lame song #1, here we go... #Eurovision This is an insult to chameleons everywhere!	None	−3	Negative
Her voice, "Mmm the flavour", #Eurovision	None	0	Positive

translated tweet text token is compared with elements in five lists (see below). For non-English tweet tokens, negation and intensifier lists are skipped, and English polarity lexicon [16] is replaced by multilingual National Research Council (NRC) emotion lexicon [23], which have been translated by the creators of the original English lexicon using Google Translate.

English polarity lexicon (Hu and Liu (HL)-lexicon [16])/multilingual NRC emotion lexicon[5] had a list of positive and negative English words with sentiment value of 1 or −1 (truncated).

Words found in the negation list[6], e.g., "not" and "rarely", reverse the polarity of the subsequent word [17].

Words found in the intensifier list (see Footnote 6), e.g., "absolutely" and "occasionally", increased or decreased the sentiment value of the subsequent word

---

[5] http://saifmohammad.com/WebPages/lexicons.html last accessed May. 4, 2020.

[6] https://github.com/cjhutto/vaderSentiment/blob/master/vaderSentiment/ vaderSentiment.py last accessed Jan. 10, 2020.

that had polarity [17]. Intensifiers in sequence had diminishing effects if the magnitude of their coefficients was less than 1.

List of emoticons (see Footnote 6) (:-), ;), :(, ;-{, ...) and emojis[7] (☺ 😠 😩 ☺ ☺, etc.) included positive and negative emoticons/emojis with a sentiment value of 1 or −1 (Table 4).

Table 4. Counts of positive, negative, and total words, emoticons, or emojis in the polarity lexicons

Polarity lexicon	Positive	Negative	Total
HL-lexicon	2 006	4 783	6 789
Multilingual NRC	2277	3267	5544
Emoticons	46	21	67
Emojis	740	128	868

Note 1. The counts for multilingual NRC lexicon represent the number of English words in the lexicon. The lexicon also includes the translations of these words by Google Translate into 103 other languages.

For each foreign language tweet, we used the Google Cloud Translate[8] to translate the text of the tweet into English. Then, we tokenized the translated tweet and applied the sentiment analysis algorithm in the same manner as for the tweets originally in English.

To apply the multilingual NRC lexicon, we tokenized the foreign language tweets and assigned a sentiment value for each emoji, emoticon, or word that was found in the multilingual lexicon. We did not analyze the morphology of the words but only looked for identical matches in the identified tweet language. Furthermore, negations and intensifiers were not considered when the multilingual lexicon was used to estimate the sentiment value of foreign language tweets.

The pre-trained and fine-tuned RoBERTa sentiment analysis method was applied only to tweets in English or tweets that were translated to English, because the English language tweets form the majority of the data and training or fine-tuning the language models for all 41 languages present was unfeasible. The distribution of languages in the analyzed tweets are shown in Table 5. The tweets with undefined language are tweets that have no clear main language, such as tweets consisting of hashtags, interjections, names, and emojis [21]. We also tried fine-tuning a multilingual DistilBERT language model[9], but RoBERTa seemed to give better results in pilot experiments.

---

[7] http://kt.ijs.si/data/Emoji_sentiment_ranking/index.html last accessed Jan. 10, 2020.

[8] https://cloud.google.com/translate/ last accessed Mar. 30, 2020.

[9] https://huggingface.co/distilbert-base-multilingual-cased.

Table 5. Distribution of languages in the sentiment-analyzed tweets

Language	Number of tweets	Language	Number of tweets
English	119788	Haitian Creole	171
Spanish	38331	Hungarian	165
German	17291	Romanian	161
French	8839	Welch	159
Italian	5513	Basque	132
Undefined	4792	Slovenian	114
Dutch	4643	Hebrew	96
Russian	1946	Czech	79
Indonesian	1141	Japanese	75
Portuguese	1139	Lithuanian	71
Polish	1034	Latvian	35
Swedish	974	Serbian	28
Catalan	810	Bulgarian	28
Greek	784	Hindi	20
Tagalog	558	Ukrainian	19
Estonian	480	Vietnamese	4
Norwegian	411	Arabic	3
Danish	364	Armenian	1
Finnish	318	Persian	1
Turkish	215	Korean	1
Icelandic	196		

The sentiment analysis dataset has 210,930 tweets by 66,631 distinct users, with an average number of tweets per user of 3.17 (lowest 1, highest 94). The sentiment analysis was done on a tweet level rather than on an author level, because viewers can cast their votes multiple times during the televoting period [11].

3.6 Tallying of Final Results

After analyzing 1,094,181 Eurovision tweets, we obtained 210,930 (19.3%) tweets to be classified as having positive, negative, or no sentiment. We discarded 73.4% of the tweets because they had no identified target or source country. In addition, we also excluded tweets from countries that were not eligible to vote in Eurovision. Therefore, both the source and target country information was important; Eurovision televoting rules prohibited viewers from voting for their own countries to prevent national pride from skewing the results. The distribution of the analyzed tweets is shown in Table 6.

A problem with predicting the ranking based only on the counts of positive or negative tweets is that some countries send many more tweets than others, while

Table 6. Distribution of analyzed tweets

Tweet filter	Tweet count	Tweets retained after filtering
Analyzed tweets	1,094,181 (100%)	100%
No Identified target country	632,450 (57.8%)	42,2%
Multiple Identified target countries	27,867 (2.5%)	39,7%
No Identified source country	170,622 (15.6%)	24,1%
Source country not eligible voter	972 (0.1%)	24%
Same source and target country	51,340 (4.7%)	19.3%
Tweets for sentiment analysis	210,930 (19.3%)	

Note 2. Each row only counts tweets which are excluded using the filters in above rows. For example, the third row shows the number of tweets which do have a single identified target country but do not have an identified source country. The percentage of total tweets is shown in parenthesis.

Table 7. Source country distribution in the sentiment-analyzed tweets

United Kingdom	83352 (39.52%)	Serbia	614 (0.29%)
Spain	34158 (16.19%)	Slovenia	562 (0.27%)
Germany	22600 (10.71%)	Belarus	489 (0.23%)
Italy	14074 (6.67%)	Hungary	477 (0.23%)
France	11826 (5.61%)	Croatia	388 (0.18%)
Netherlands	7508 (3.56%)	Iceland	332 (0.16%)
Russia	6219 (2.95%)	Czech Republic	276 (0.13%)
Australia	4474 (2.12%)	Cyprus	227 (0.11%)
Belgium	4044 (1.92%)	Estonia	215 (0.10%)
Sweden	2749 (1.30%)	Albania	201 (0.10%)
Switzerland	2020 (0.96%)	Lithuania	198 (0.09%)
Portugal	1893 (0.90%)	North Macedonia	123 (0.06%)
Poland	1658 (0.79%)	Malta	116 (0.05%)
Greece	1549 (0.73%)	Latvia	88 (0.04%)
Finland	1472 (0.70%)	Montenegro	76 (0.04%)
Norway	1375 (0.65%)	Azerbaijan	56 (0.03%)
Austria	1290 (0.61%)	Georgia	54 (0.03%)
Denmark	1287 (0.61%)	Armenia	12 (0.01%)
Romania	1068 (0.51%)	San Marino	11 (0.01%)
Israel	923 (0.44%)	Moldova	9 (0.00%)
Ireland	867 (0.41%)		

in the competition all countries give out the same number of televoting points. Indeed, 39.5% of all 210,930 tweets, or 62.7% of English and undefined language tweets, are from the United Kingdom. The tweet source country distribution is shown in Table 7. This makes the count-based prediction results biased towards the opinions of voters in the UK and other countries with many tweets, as they are over-represented in the dataset.

To eliminate this bias, we experimented with applying televoting rules to the tweets, such that each country gives out the same amount of points based on the tweets from that country. We thus predicted the ranking of each country using four different methods:

1. The total number of positive tweets for each country is counted
2. The total number of positive - negative tweets for each country is counted
3. Televoting rules applied to the positive tweets
4. Televoting rules applied to the positive - negative tweets

The televoting rules are as follows:

The viewers from participating countries could cast votes for finalist countries via telephone, SMS, or app [11]. For each voting country, the votes from that country were used to award points to the finalist countries, ranked from those with the most votes to those with the least votes. The country with the most votes received 12 points, that with the second most votes received 10 points, and the countries ranked from 3 to 10 received a decreasing number of points from 8 to 1, respectively. In addition, the juries awarded points in the same fashion, ranking countries according to preference. The final score for each finalist was the sum of the televoting and jury points from each country and jury.

4   Experimental Results

4.1   Televoting Algorithm

Table 9 shows the predicted ranks calculated with the four methods from Sect. 3.6 and their correlation with the actual televoting ranks in the Eurovision final. A higher correlation corresponds to a better prediction. These results use RoBERTa for sentiment analysis using only English and undefined language tweets. The different sentiment analysis methods are compared in Sect. 4.3. The televoting algorithm provided better predictions than a simple sentiment analysis algorithm in Table 9.

4.2   Different Sampling Windows

Table 10 shows the Spearman's rank correlation coefficients (ρ), and the Kendall's rank correlation coefficients (τ) for each of the three time windows in Fig. 2 (performances, televoting, and both) using the televoting algorithm with different sentiment analysis methods. These values measure the similarity between the original ranks and the predicted ranks. The televoting period (21:10 GMT to 22:08 GMT) was the optimal time window for sampling tweets to predict ranking using tweet sentiment analysis. One reason for this could be that this was the period when the voting actually happened, and people would then tweet more about the artists they really voted for. When the tweets from the televoting period were analyzed using RoBERTa, the Spearman correlation (ρ) between the predicted score and actual score was better if all negative sentiment tweets were counted as negative votes.

4.3 Sentiment Analysis

Table 10 shows that the use of the sentiment analysis methods increased the correlation in almost all situations. The greatest increases occurred in tweets sent during the performance period. When tweets were used from only the televoting period, during which the number of tweets was the lowest, both English-translated and multilingual sentiment analyses yielded lower correlations. The best results were achieved by using RoBERTa sentiment analysis for English and undefined language tweets. This method had a Spearman's correlation (ρ) of 0.74 for the country ranking prediction based on tweets during the televoting period.

4.4 Prediction Error

Explanatory variables in Table 11 include three variables that aim to capture the prediction error: *country population over 40 million, average singer(s) age under 30* and *order of appearance*. The p-value for *country population over 40 million* from televoting and performance + televoting period is less than the alpha level of 0.05, which indicates that it is statistically significant. While tweets that were identified as having the same target and source country were rejected, the tweets from highly populated countries could have provoked conversation in other nations (Table 11).

The singer's average age under 30 decreased the prediction error, but it wasn't statistically significant with p-values of 0.83 and 0.62 in Table 11. The order of appearance effect was significantly lower during the televoting period than during the performances + televoting period as the variable was increased from 0.12 to 0.29 and the p-value decreased from 0.31 to 0.06. This effect can be explained by the low Twitter activity at the beginning of the program, as shown in Fig. 2.

The population size effect could explain why Spain, France, and the United Kingdom, which were widely discussed by English-speaking Twitter users and therefore were predicted to be higher rank, received only a few points in the actual televoting (see Fig. 5). Likely due to ordering and population size effects, Albania and Denmark, which were almost ignored on Twitter, received a fair number of votes in the actual televoting.

4.5 Non-English Tweets

This paper analyzed two methods to determine the sentiment of non-English tweets: whether it was better to translate the original text to English and analyze the English sentiments, or to translate the lexicon from English to the original target language. Table 8 shows the statistics for these methods after the tweets were analyzed.

Surprisingly, it turns out that translating the tweets or using a multilingual lexicon for the sentiment analysis did not improve the correlation between the predicted and actual votes (Table 10). This could be attributed to mistranslation [29], cultural differences, and different sense distributions [22]. It is worth

Table 8. Sentiment analysis tweet statistics for the different sentiment analysis methods

Method	Positive tweets	Negative tweets	No sentiment identified	Foreign language tweets
Only English + undefined Tweets	52,568 (24.9%)	16,737 (7.9 %)	55,275 (26.2%)	86,350 (40.9%)
English + Google translated Tweets	85,472 (40.5%)	28,505 (13.5%)	96,953 (46.0%)	–
Multilanguage polarity lexicons	77,713 (36.8%)	26,112 (12.4 %)	107,105 (50.8%)	–
Only English + undefined Tweets RoBERTa	79,998 (37.9%)	44,582 (21.1%)	–	86,350 (40.9%)
English + Google translated Tweets RoBERTa	135,368 (64.2%)	75,562 (35.8%)	–	–

considering that even when using either translated tweetset or multilingual sentiment analysis, which both include also the English sentiment analysis, 59.1% or 52.1% respectively of the tweets (sent during performance and televoting period) voting were originally in English + undefined Tweets as can be calculated from Table 10.

4.6 'Voting Blocs' in Eurovision

The televoters may vote not for the best performance but rather for neighboring or favorite European countries or particular ethnicities. Some studies have identified "voting blocs" in Eurovision [6,12,13,26,31,32]. We investigated similar trends in televote ranking based on tweets. Russia received high points from the countries with a high ethnically Russian population, such as Belarus, Latvia, and Estonia (Fig. 4). Fenn et al. [12] found that Cyprus and Greece often exchanged votes. A similar tendency is shown in Fig. 4. The countries in the Nordic bloc (Norway, Sweden, Iceland, Denmark, and Finland) all gave 12 points to Iceland and high points also to Norway. Facebook, Twitter, Instagram, and YouTube dominate the social media world in most countries, but Russia (and some surrounding areas, such as Latvia and Estonia) often prefer regional replacements for these platforms, e.g.. VKontakte (VK). This and possible political considerations might have led to Russia receiving a much higher number of points in the actual contest than in Twitter-based prediction.

Younger generations are not interested in immigrant/ethnic roots or do not express support for their home countries on social media. Particularly in countries with a high ethnically Russian population, this "blocking" behavior combined with the low number of Twitter users [27] may have resulted in a very poor prediction for Russia.

5 Discussion

Most analyses that are based on data from different social media depend on access to samples. The data is typically provided through an API, which is an unavoidable "black box" between the data source and researchers [4,9,28]. Moreover, data sampling of the Twitter API has been shown to be biased [28,28], which imposes additional limitations on any analysis based on Twitter data.

Twitter users are younger, better educated, and more liberal [1,10,14,19, 20,33,34]. Furthermore, among Twitter users, rates of tweeting are extremely skewed [3]. The liberal bias may have led to underscoring for Russia and overscoring for France and Iceland in Table 9. Russia's underscoring could be attributed to political reasons and France's overscoring to the singer's background in the French LGBT movement. Figure 3 shows that Iceland's performances received a lot of positive and negative tweets from other countries. Iceland's self-proclaimed "techno-dystopian" band was predicted to be top-ranked, possibly because its deviation from the Eurovision's usually pop-centered performances courted controversy. The predicted televotes for Estonia may have been influenced by the tweets regarding Iceland's performance; therefore, Estonia's predicted televotes were higher than its actual televotes. Israel received substantial attention from Twitter users, as shown in Table 9, likely due to the fact that the host country puts significant effort into social media promotion.

It can be seen in Fig. 5 that the predicted televote scores given by United Kingdom do not differ very much from their actual given scores. The sum of absolute voting errors for the United Kingdom is 34, while the average sum of absolute voting errors is 51.98. Moldova, Armenia, and San Marino were not taken into account when calculating this average, because their predicted scores contain multiple instances of same points due to very small number of English tweets (7, 5, and 4, respectively). In addition, 62.7% of the analyzed English and undefined language tweets came from the United Kingdom. This suggests that a larger number of tweets from all countries, and more accurate multilingual sentiment analysis would notably increase the results' accuracy.

6   Conclusion

As the results above show, Twitter tweets have fairly strong correlation with televoting behaviour. However, users of Twitter and other social media are not demographically representative of the population. Nevertheless, with appropriate adjustments, e.g. liberal bias, ordering, big country and age effect and a larger data set, social media posts could be a helpful data source for obtaining tangible predictions. Televoting ranking prediction accuracy does not benefit from translation of non-English tweets or using automatic translations of existing English lexicons for non-English tweets. Automatic translation may not convey the same sentiment because of poor translation quality or because the word and its translation are used differently in different languages.

Two main areas of improvement and further work would be increasing the number of analyzed tweets and more sophisticated sentiment analysis that does not rely on machine translation of the tweets. While our preliminary experiments with multilingual BERT were not as promising as with RoBERTa, fine-tuning sentiment analysis models for other languages in addition to English could make the sentiment analysis more accurate, as less tweets would have to be translated or discarded. The votes of the United Kingdom were predicted accurately likely due to the large number of tweets. The analysis could also be expanded to other platforms which would certainly improve results from for example Russia. The identification of the source country proved to be relatively difficult and can possibly be improved by for example taking into account the tweet text. Additionally, the target country identification could possibly use the tweet time to determine the target country or include additional methods for determining a sentiment towards multiple target countries within a single tweet.

Acknowledgments. Thanks to Dimitri Demergis for sharing his data.

A   Dictionary

Finalist country: Countries that advanced to the finals of Eurovision 2019
Geolocation: Satellite based location, included in the properties of the tweet from twitter API
Hashflag: Hashtags such as #NOR or #SWE
Negative tweets: Tweets with negative sentiment
Participating country: Countries that participated in Eurovision 2019
Polarity The sign of the sentiment value; that is, if the sentiment value was positive, the polarity was positive and vice versa for negative sentiments
Polarity lexicon: A list of words with polarity
Positive tweet: Tweets with positive sentiment
Positive - negative tweets: The value obtained from the subtraction of negative tweets from the positive tweets
Source country: The country from which a tweet was sent, obtained by checking the tweet's geolocation

Target country: The country that a tweet voted for, obtained by checking whether the tweet text contained a hashflag or a keyword

Televoting rules/Televoting algorithm: The scoring system of Eurovision; that is, how the points are awarded (each participating country gives a certain amount of points to a finalist country)

Undefined language tweets: Tweets with no clear main language, such as tweets consisting of hashtags, interjections, names, and emojis.

Table 9. Actual televote ranking and predicted rankings

Country	Actual televote rank	Positive tweets pred. rank by number of tweets	Positive - negative tweets pred. rank by number of tweets	Positive tweets pred. rank by televoting algorithm	Positive - negative tweets pred. rank by televoting algorithm
Norway	1	3	2	4	3
Netherlands	2	12	11	8	8
Italy	3	4	6	2	2
Russia	4	16	9	12	9
Switzerland	5	6	5	6	7
Iceland	6	1	1	1	1
Australia	7	2	3	3	4
Azerbaijan	8	7	8	10	10
Sweden	9	18	12	22	17
San Marino	10	21	25	19	21
Slovenia	11	19	26	20	19
North Macedonia	12	22	18	17	18
Serbia	13	14	14	14	11
Spain	14	8	7	7	6
Denmark	15	20	15	24	22
Estonia	16	10	10	13	13
Albania	17	25	20	24	25
France	18	5	4	5	5
Israel	19	9	17	11	14
Cyprus	20	15	13	16	16
Greece	21	11	24	15	15
Malta	22	23	19	24	24
Belarus	23	13	16	21	23
Czech Republic	24	26	22	18	20
United Kingdom	25	17	21	8	12
Germany	26	24	23	24	25
Spearman correlation coefficient (ρ)	1.0	0.523	0.627	0.545	0.649

Table 10. Ranking correlation for predicting by televoting algorithm in different time windows

Method	Sampling period	Tweets with identified target and source country, and sentiment (if applicable)	Spearman correlation coefficient (ρ)	Kendall rank correlation coefficient (τ)
No sentiment analysis, tweet counter as vote	Performances + televoting	210,930	0.42	0.30
	During performances	183,733	0.34	0.24
	During televoting	27,208	0.59	0.44
Only English + undefined tweets	Performances + televoting	69,291	0.51 (0.49)	0.37 (0.34)
	During performances	61,670	0.46 (0.49)	0.32 (0.33)
	During televoting	7,625	0.60 (0.62)	0.46 (0.48)
English + Google translate	Performances + televoting	113,946	0.49 (0.47)	0.34 (0.33)
	During performances	99,936	0.45 (0.42)	0.33 (0.29)
	During televoting	14,015	0.57 (0.57)	0.44 (0.45)
Multilanguage polarity lexicons	Performances + televoting	103,794	0.49 (0.50)	0.35 (0.35)
	During performances	91,152	0.46 (0.44)	0.33 (0.32)
	During televoting	12,647	0.56 (0.60)	0.43 (0.46)
Only English + undefined RoBERTa	Performances + televoting	124,580	0.65 (0.54)	0.47 (0.37)
	During performances	110,404	0.62 (0.47)	0.44 (0.33)
	During televoting	14,184	0.74 (0.71)	0.55 (0.53)
English + Google translate RoBERTa	Performances + televoting	210,930	0.58 (0.50)	0.42 (0.35)
	During performances	183,733	0.52 (0.44)	0.38 (0.33)
	During televoting	27,208	0.72 (0.63)	0.55 (0.50)

Note 3. Only positive tweets in the brackets. Performances + televoting includes 1-min break after performances.

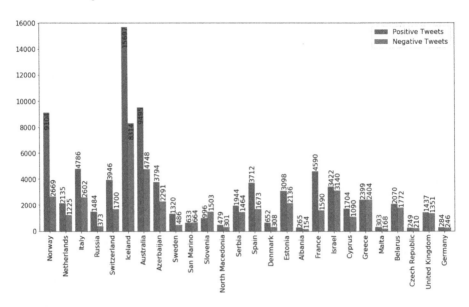

Fig. 3. Number of positive and negative tweets in English or undefined language received by each participating country. The tweets were classified as positive or negative by using RoBERTa sentiment analysis.

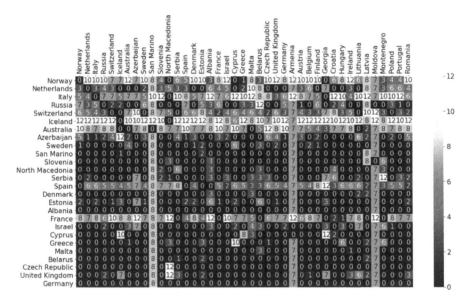

Fig. 4. Predicted Scoreboard. *Note 4.* X-axis: prediction results. Y-axis: contestants. Predicted televote scores by countries from positive - negative tweets. Some countries, such as San Marino, awarded the same number of points to many countries because countries with a shared third rank all received 8 points. The ranks were shared since there were not enough tweets from San Marino to rank the countries.

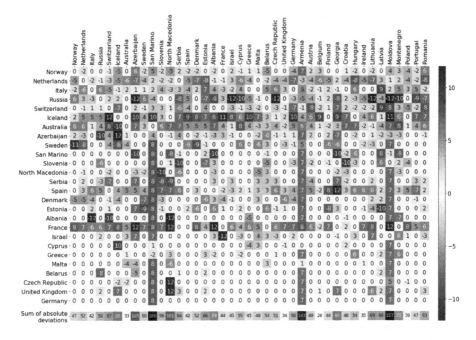

Fig. 5. The difference between the predicted score and the actual televotes. *Note 5.* Added at the bottom is also the sum of absolute deviations by country. This gives a general overview of how well the points given by a single country were predicted

Table 11. Prediction error

Country	Order of appearance	Population > 40 Million	Performers age avg. < 30	Prediction error performances + televoting period	Prediction error televoting period
Malta	1	0	1	−2	−1
Albania	2	0	0	−8	2
Czech Republic	3	0	0	4	−2
Germany	4	1	1	1	2
Russia	5	0	0	−5	−1
Denmark	6	0	1	−7	2
San Marino	7	0	0	−11	−9
North Macedonia	8	0	0	-6	−5
Sweden	9	0	0	−8	−1
Slovenia	10	0	1	−8	−8
Cyprus	11	0	0	4	3
Netherlands	12	0	1	−6	−4
Greece	13	0	1	6	−3
Israel	14	0	1	5	−2
Norway	15	0	0	−2	−5
United Kingdom	16	1	1	13	13

(continued)

Table 11. (*continued*)

Country	Order of appearance	Population > 40 Million	Performers age avg. < 30	Prediction error performances + televoting period	Prediction error televoting period
Iceland	17	0	1	5	5
Estonia	18	0	1	3	−1
Belarus	19	0	1	0	2
Azerbaijan	20	0	1	−2	−3
France	21	1	1	13	10
Italy	22	1	1	1	0
Serbia	23	0	1	2	0
Switzerland	24	0	1	−2	−4
Australia	25	0	0	3	3
Spain	26	1	1	8	12

- Y is the prediction error
- X is the running number ($p = 0.059$)
- $S = 1$ if the average age of the performers is <30, otherwise $S = 0$ ($p = 0.620$)
- $C = 1$, if the country has population of >40 million, otherwise $C = 0$ ($p = 0.020$)
- The correlation coefficient R^2 is 0.448

The regression line with prediction error (Televote rank - Tweets Predicted Rank) calculated from tweets from the televoting period: $Y = -2.85 + 0.125X + 8.434C - -0.415S$, Y is prediction error, where

- X is the running number ($p = 0.311$)
- $S = 1$ if the average age of the performers is <30, otherwise $S = 0$ ($p = 0.834$)
- $C = 1$, if the country has population of >40 million, otherwise $C = 0$ ($p = 0.001$)
- R^2 is 0.468

Russia is not marked as having a population over 40 million because of the low number of Twitter users[27].

References

1. Barberá, P., Rivero, G.: Understanding the political representativeness of Twitter users. Soc. Sci. Comput. Rev. **33**(6), 712–729 (2015)
2. Blangiardo, M., Baio, G.: Evidence of bias in the Eurovision song contest: modelling the votes using Bayesian hierarchical models. J. Appl. Stat. **41**(10), 2312–2322 (2014)
3. Blank, G.: The digital divide among twitter users and its implications for social research. Soc. Sci. Comput. Rev. **35**(6), 679–697 (2017)
4. Bruns, A., Stieglitz, S.: Quantitative approaches to comparing communication patterns on Twitter. J. Technol. Hum. Serv. **30**(3–4), 160–185 (2012)

5. Clerides, S., Stengos, T., et al.: Love thy neighbour, love thy kin: Strategy and bias in the Eurovision song contest. Ekonomia **15**(1), 22–44 (2012)
6. Dekker, A.: The Eurovision song contest as a 'friendship' network. Connections **27**(3), 53–58 (2007)
7. Demergis, D.: Predicting Eurovision song contest results by interpreting the tweets of Eurovision fans. In: 2019 Sixth International Conference on Social Networks Analysis, Management and Security (SNAMS), pp. 521–528. IEEE (2019)
8. Devlin, J., Chang, M.W., Lee, K., Toutanova, K.: BERT: pre-training of deep bidirectional transformers for language understanding. arXiv preprint arXiv:1810.04805 (2018)
9. Driscoll, K., Walker, S.: Big data, big questions| working within a black box: transparency in the collection and production of big twitter data. Int. J. Commun. **8**, 20 (2014)
10. Duggan, M.: Mobile messaging and social media 2015. Pew Research Center (2015). https://www.pewresearch.org/internet/2015/08/19/mobile-messaging-and-social-media-2015/. Accessed 09 Mar 2020
11. European Broadcasting Union: Eurovision Song Contest - Voting. https://eurovision.tv/about/voting. Accessed 09 Mar 2020
12. Fenn, D., Suleman, O., Efstathiou, J., Johnson, N.F.: How does Europe make its mind up? Connections, cliques, and compatibility between countries in the Eurovision song contest. Phys. A Stat. Mech. Appl. **360**(2), 576–598 (2006)
13. Gatherer, D.: Comparison of Eurovision song contest simulation with actual results reveals shifting patterns of collusive voting alliances. J. Artif. Soc. Soc. Simul. **9**(2), 1–13 (2006)
14. Greenwood, S., Perrin, A., Duggan, M.: Social media update 2016. Pew Res. Cent. **11**(2), 1–10 (2016)
15. Haan, M.A., Dijkstra, S.G., Dijkstra, P.T.: Expert judgment versus public opinion-evidence from the Eurovision song contest. J. Cult. Econ. **29**(1), 59–78 (2005)
16. Hu, M., Liu, B.: Mining and summarizing customer reviews. In: Proceedings of the Tenth ACM SIGKDD International Conference on Knowledge Discovery and Data Mining, pp. 168–177. ACM (2004)
17. Hutto, C.J., Gilbert, E.: VADER: a parsimonious rule-based model for sentiment analysis of social media text. In: Eighth International AAAI Conference on Weblogs and Social Media. The AAAI Press (2014)
18. Liu, Y., et al.: RoBERTa: a robustly optimized BERT pretraining approach (2019)
19. Mellon, J., Prosser, C.: Twitter and facebook are not representative of the general population: political attitudes and demographics of British social media users. Res. Polit. **4**(3), 1–9 (2017)
20. Mislove, A., Lehmann, S., Ahn, Y.Y., Onnela, J.P., Rosenquist, J.N.: Understanding the demographics of Twitter users. In: Fifth International AAAI Conference on Weblogs and Social Media. The AAAI Press (2011)
21. Mitja, T.: Evaluating language identification performance (2020). https://blog.twitter.com/engineering/en_us/a/2015/evaluating-language-identification-performance.html
22. Mohammad, S.M., Salameh, M., Kiritchenko, S.: How translation alters sentiment. J. Artif. Intell. Res. **55**, 95–130 (2016)
23. Mohammad, S.M., Turney, P.D.: Crowdsourcing a word-emotion association lexicon. Comput. Intell. **29**(3), 436–465 (2013)
24. Ochoa, A., Hernández, A., González, S., Jöns, S., Padilla, A.: Hybrid system to determine the ranking of a returning participant in Eurovision. In: 2008 Eighth International Conference on Hybrid Intelligent Systems, pp. 489–494. IEEE (2008)

25. Ochoa, A., Hernández, A., Sánchez, J., Muñoz-Zavala, A., Ponce, J.: Determining the ranking of a new participant in Eurovision using cultural algorithms and data mining. In: 18th International Conference on Electronics, Communications and Computers (CONIELECOMP 2008), pp. 47–52. IEEE (2008)
26. Orgaz, G.B., Cajias, R., Camacho, D.: A study on the impact of crowd-based voting schemes in the 'Eurovision' European contest. In: Proceedings of the International Conference on Web Intelligence, Mining and Semantics, pp. 1–9. Association for Computing Machinery (2011)
27. Longley, P.A., Adnan, M., Lansley, G.: The geotemporal demographics of Twitter usage. Environ. Plan. **47**, 465–484 (2015)
28. Pfeffer, J., Mayer, K., Morstatter, F.: Tampering with Twitter's sample API. EPJ Data Sci. **7**(1), 50 (2018)
29. Lohar, P., Afli, H., Way, A.: Maintaining sentiment polarity in translation of user-generated content. Prague Bull. Math. Linguist. **108**(1), 73–84 (2017)
30. Socher, R., et al.: Recursive deep models for semantic compositionality over a sentiment treebank. In: Proceedings of the 2013 Conference on Empirical Methods in Natural Language Processing, pp. 1631–1642 (2013)
31. Spierdijk, L., Vellekoop, M.: The structure of bias in peer voting systems: lessons from the Eurovision song contest. Empir. Econ. **36**(2), 403–425 (2009)
32. Stockemer, D., Blais, A., Kostelka, F., Chhim, C.: Voting in the Eurovision song contest. Politics **38**(4), 428–442 (2018)
33. Vaccari, C., et al.: Social media and political communication. A survey of Twitter users during the 2013 Italian general election. Rivista italiana di scienza politica **43**(3), 381–410 (2013)
34. Wojcik, S., Hughes, A.: Sizing up Twitter users. Pew Research Center. https://www.pewinternet.org/2019/04/24/sizing-up-twitter-users (2019). Accessed 09 Mar 2020

Improving Results on Russian Sentiment Datasets

Anton Golubev[1]([⊠])(ⓘ) and Natalia Loukachevitch[2]([⊠])(ⓘ)

[1] Bauman Moscow State Technical University, Moscow, Russia
antongolubev5@yandex.ru
[2] Lomonosov Moscow State University, Moscow, Russia
louk_nat@mail.ru

Abstract. In this study, we test standard neural network architectures (CNN, LSTM, BiLSTM) and recently appeared BERT architectures on previous Russian sentiment evaluation datasets. We compare two variants of Russian BERT and show that for all sentiment tasks in this study the conversational variant of Russian BERT performs better. The best results were achieved by BERT-NLI model, which treats sentiment classification tasks as a natural language inference task. On one of the datasets, this model practically achieves the human level.

Keywords: Targeted sentiment analysis · BERT · Natural language inference

1 Introduction

Sentiment analysis studies are currently based on the application of deep learning approaches, which requires training and testing on specialized datasets. For English, popular sentiment analysis datasets include: Stanford Sentiment Treebank datasets SST [19], IMDB dataset of movie reviews [12], Twitter sentiment datasets [13,15], and many others. For other languages, much less datasets have been created. In Russian several sentiment evaluations were previously organized, including ROMIP2012–2013 and SentiRuEval2015–2016 [4,9,10], which included the preparation of annotated data on reviews (movies, books and digital cameras), news quotes, and Twitter messages. The best results on these datasets were obtained with classical machine learning techniques such as SVM [4], early neural network approaches [2], or even engineering methods based on rules and lexicons [8]. Currently, the results achieved in the above-mentioned Russian evaluations can undoubtedly be improved.

In this study, we test standard neural network architectures (CNN, LSTM, BiLSTM) and recently appeared BERT architectures on previous Russian sentiment evaluation datasets. We compare two variants of Russian BERT [6] and show that for all sentiment tasks in this study the conversational variant of Russian BERT performs better. The best results were achieved by BERT-NLI model, which treats sentiment classification problem as the natural inference task.

A. Filchenkov et al. (Eds.): AINL 2020, CCIS 1292, pp. 109–121, 2020.
https://doi.org/10.1007/978-3-030-59082-6_8

In one of the tasks this model practically achieves the human level of sentiment analysis.

The contributions of this paper are as follows:

- we renew previous results on five Russian sentiment analysis datasets using the state-of-the-art methods,
- we test new conversational Russian BERT model in several sentiment analysis tasks and show that it is better than previous Russian RuBERT model,
- we show that the BERT model, which treats sentiment analysis as a natural language inference task achieves the best results on all datasets under analysis.

This paper is structured as follows. In Sect. 2 we present sentiment analysis datasets previously created for Russian shared tasks, best methods and achieved results in previous evaluations. Section 3 describes preprocessing steps and methods applied to sentiment analysis tasks in the current study, including several BERT-based models. Section 4 presents the achieved results. In Sect. 5 we analyse the errors of models on difficult examples. Section 6 describes other available Russian sentiment analysis datasets and methods applied to these datasets.

2   Datasets

In our study we consider five Russian datasets annotated for previous Russian sentiment evaluations: news quotes of the ROMIP-2013 evaluations [4] and Twitter datasets of two SentiRuEval evaluations 2015–2016 [9,10]. Table 1 presents the datasets under evaluation, the volumes of their training and test parts, main quality measures, achieved results, and the best methods. Table 2 contains the distribution of the datasets texts by sentiment classes.

Table 1. Datasets under evaluation.

Dataset	Train vol	Test vol	Metrics	Result	Method
News Quotes ROMIP-2013[a]	4260	5500	$F_1\ macro$	62.1	Lexicons+Rules
SentiRuEval-2015 Telecom[b]	5000	5322	$F_1^{+-}\ macro$	50.3	SVM
SentiRuEval-2015 Banks[b]	5000	5296	$F_1^{+-}\ macro$	36.0	SVM
SentiRuEval-2016 Telecom[c]	8643	2247	$F_1^{+-}\ macro$	55.9	2-layer GRU
SentiRuEval-2016 Banks[c]	9392	3313	$F_1^{+-}\ macro$	55.1	2-layer GRU

[a] http://romip.ru/en/collections/sentiment-news-collection-2012.html
[b] https://drive.google.com/drive/folders/1bAxIDjVz_0UQn-iJwhnUwngjivS2kfM3
[c] https://drive.google.com/drive/folders/0BxlA8wH3PTUfV1F1UTBwVTJPd3c

2.1   News Quotes Dataset

For creating news quotes collection, opinions in direct or indirect speeches were extracted from news articles [4]. The task was to classify quotations as neutral, positive or negative speaker comment about the topic of the quotation. It can be seen in Table 2 that class distribution in the dataset was rather balanced. The main quality measure was F_1 *macro*.

The participants experimented with classical machine learning approaches such as Naive Bayes and SVM classifiers, but the best results were obtained by a knowledge-based approach using a large sentiment lexicon and rules: 62.1 of F_1 measure and 61.6 of accuracy score. This can be explained with great variety of topics and topic-related sentiment discussed in news quotes [4].

Table 2. Class distribution by datasets (%).

Dataset	Train sample			Test sample		
	Positive	Negative	Neutral	Positive	Negative	Neutral
News Quotes ROMIP-2013	16	36	48	11	33	56
SentiRuEval-2015 Telecom	19	32	49	10	23	67
SentiRuEval-2015 Banks	7	34	59	8	15	79
SentiRuEval-2016 Telecom	15	29	56	10	46	44
SentiRuEval-2016 Banks	8	18	74	10	22	68

2.2   Twitter Datasets

Twitter datasets were annotated for the task of reputation monitoring [1,9]. The goal of Twitter sentiment analysis at SentiRuEval was to find sentiment-oriented opinions or positive and negative facts about two types of organizations: banks and telecom companies. In such a way the task can be classified as targeted (entity-oriented) sentiment analysis problem. Similar evaluations were organized twice in 2015 and 2016, during which four target-oriented datasets were annotated (Table 1). In 2016 training datasets in both domains were constructed by uniting of training and test data of the 2015 evaluation and, therefore they were much larger in size [10].

The participating systems were required to perform a three-way classification of tweets: positive, negative or neutral. It can be seen in Table 2 that neutral class was prevailing in the datasets. For this reason, the main quality measure was F_1^{+-}*macro* measure, which was calculated as the average value between F_1 measure of the positive class and F_1 measure of the negative class. F_1 measure of the neutral class was ignored because this category is usually not interesting to know. But this does not reduce the task to the two-class prediction because erroneous labeling of neutral tweets negatively influences on F_1^+ and F_1^-. Additionally micro-average F_1^{+-}*micro* measures were calculated for two sentiment classes [9,10].

It can be seen in Table 1 that the results in 2016 are much higher than in 2015 for the same tasks. There can be two reasons for this. The first reason is the larger volume of the training data in 2016. The second reason is the use by the participants of more advanced methods, including neural network models and embeddings.

3   Methods

We compare the following groups of sentiment analysis methods on the above-described datasets. SVM with pre-trained embeddings is a baseline for our study. We chose FastText[1] embeddings (dimension 300) because of its better results compared with other types of Russian embeddings such as ELMo (see Footnote 1), Word2Vec[2], and GloVe[3] in preliminary studies. To submit data to the SVM algorithm, averaging token embeddings in a sentence was used. Grid search mechanism from scikit-learn[4] framework was utilized to obtain optimal hyper-parameters.

3.1   Preprocessing

Since most of the data are tweets containing noise information, significant text preprocessing was implemented. The full cycle contained the following steps:

- lowercase cast;
- replacing URLs with *url* token;
- replacing mentions with *user* token;
- replacing hashtags with *hashtag* token;
- replacing emails with *email* token;
- replacing phone numbers with *phone* token;
- replacing emoticons with appropriate tokens like *sad, happy, neutral*;
- removing all special symbols except punctuation marks;
- replacing any repeated more than 2 times in a row letter with 2 repetitions of that letter;
- lemmatization and removing stop words.

It is worth to note that the last point was applied only in the case of SVM and classic neural networks. For BERT-based methods it did not make any difference and gave not a considerable change of about 0.01%.

---

[1] http://docs.deeppavlov.ai/en/master/features/pretrained_vectors.html.
[2] https://rusvectores.org/ru/models/.
[3] http://www.cs.cmu.edu/~afm/projects/multilingual_embeddings.html.
[4] https://scikit-learn.org/stable/.

3.2 Classical Neural Networks

The first group of methods is a set of classical convolutional and LSTM neural networks.

The architecture of the convolutional neural network considered in this paper is based on approaches [5,22]. Input data is represented as a matrix of size $s \times d$, where s is the number of tokens in the tweet and d is the dimension of the embedding space. The optimal matrix height $s = 50$ was chosen experimentally. If necessary, a sentence is truncated or zero-padded to the required length.

After that several convolution operations of various sizes are applied to this matrix in parallel. A single branch of convolution involves a filtering matrix $w \in \mathbb{R}^{h \times d}$ with the size of the convolution h equal the number of words it covers. Then output of each branch is max-pooled. This helps to extract the most important information for each convolution, regardless of feature position in the text. After all convolution operations, obtained vectors are concatenated and sent to a fully connected layer, which is then passed through the softmax layer to give the final classification probabilities. In our model we chose the number of convolution branches equal to 4 with windows sizes $(2, 3, 4, 5)$ respectively. To reduce overfitting, dropout layers with probability of $p = 0.5$ were added after max-pooling and fully connected layers.

The main idea of LSTM recurrent networks is the introduction of a cell state of dimension $c_t \in \mathbb{R}^d$ equal to the dimension of network, running straight down the entire chain and ability of LSTM to remove or add information to the cell state using special structures called gates. This helps to avoid the exploding and vanishing gradient problems during the backpropagation training stage in recurrent neutral networks. In our work, we chose d equal to the size of token embeddings.

Besides, we used the bidirectional LSTM (BiLSTM) model, which represents two LSTMs stacked together. Two networks read the sentence from different directions and their cell states are concatenated to obtain vector of dimension $c_t \in \mathbb{R}^{2d}$. As well as in LSTM network, this vector is sent to a fully connected layer of size 40 and then passed through a softmax layer to give the final classification probabilities.

In both LSTM architectures we used dropout to reduce over-fitting by adding a dropout layer with probability of $p = 0.5$ before and after the fully connected layers. For all described neural networks we used pre-trained Russian FastText embeddings with dimension of $d = 300$.

3.3 Fine-Tuning BERT Model

The second group of methods is based on two pre-trained Russian BERT models and several approaches of application of BERT [6] to the sentiment analysis task. The utilized approaches can be subdivided into single sentence classification and constructing auxiliary sentences approach [20], which converts a sentiment analysis task into a sentence-pair classification problem. It seems possible since input representation of BERT can represent both a single sentence and a pair of sentences considering them as a next sentence prediction task.

The BERT sentence-single model uses only an initial sentence as an input and represents a vanilla BERT model with an additional single linear layer with matrix $W \in \mathbb{R}^{K \times H}$ on the top. Here K denotes the number of classes and H the dimension of the hidden state. For the classification task, the first word of the input sequence is identified with a unique token *[CLS]*. The input representation is constructed by summing the initial token, segment, and position embeddings for any token in the sequence. Classification probabilities distribution is calculated using the softmax function.

The BERT sentence-pair model architecture has some differences. The input representation converts a pair of sentences in one sequence of tokens inserting special token *[SEP]* between them. The classification layer is added over the final hidden state of the first token $C \in \mathbb{R}^{H}$.

For the targeted task, there is a label for each object of sentiment analysis in a sentence so the real name of an entity was replaced by a special token. For example, the initial tweet *"Sberbank is a safe place where you can keep your savings"* is converted to *"MASK is a safe place where you can keep your savings"*.

Two sentence-pair models use auxiliary sentences and based on question answering (QA) and natural language inference (NLI) tasks. The auxiliary sentences for the targeted analysis are as follows:

- pair-NLI: *"The sentiment polarity of MASK is"*
- pair-QA: *"What do you think about MASK?"*

The answer is supposed to be one from the *Positive, Negative, Neutral* set.

In case of the general sentiment analysis task, there is one label per sentence and no objects of sentiment analysis to mask. So we proposed to assign the token to the whole sentence. Therefore the initial sentence *"56% of Rambler Group was sold to Sberbank"* is converted to *"MASK = 56% of Rambler Group was sold to Sberbank"*. The same auxiliary sentences were constructed for this task.

In our study, we compare two different pre-trained BERT models from Deep-Pavlov framework [3]:

- RuBERT, Russian, cased, 12-layer, 768-hidden, 12-heads, 180M parameters, trained on the Russian part of Wikipedia and news data[5].
- Conversational RuBERT, Russian, cased, 12-layer, 768-hidden, 12-heads, 180M parameters, trained on OpenSubtitles, Dirty, Pikabu, and Social Media segment of Taiga corpus (see Footnote 5).

During the fine-tuning procedure, we set dropout probability at 0.1, number of epochs at 5, initial learning rate at $2e - 5$, and batch size at 12.

---

[5] http://docs.deeppavlov.ai/en/master/features/models/bert.html.

4   Results

To compare different models, we calculated standard metrics such as accuracy and F_1 macro. Besides, we calculated the metrics necessary for comparison with the participants of the competition: F_1^{+-} macro and F_1^{+-} micro, which take into account only positive and negative classes. All the reported results were obtained by averaging over five runs. To distinguish two pre-trained BERT models, special label (C) is used for Conversational RuBERT.

Table 3. Results on News Quotes Dataset.

Model	Accuracy	F_1 macro	F_1^{+-} macro	F_1^{+-} micro
ROMIP-2013 [4]	61.60	62.10	–	–
SVM	69.12	61.63	74.82	75.07
CNN	68.57	60.43	73.51	74.55
LSTM	73.61	62.31	77.02	78.20
BiLSTM	74.14	62.78	77.61	78.94
BERT-single	78.90	68.07	84.33	84.45
BERT-pair-QA	79.06	68.54	84.33	84.45
BERT-pair-NLI	79.68	69.45	84.96	85.08
BERT-single (C)	79.81	**71.12**	85.05	85.10
BERT-pair-QA (C)	78.95	70.16	84.71	84.83
BERT-pair-NLI (C)	**80.28**	70.62	**85.52**	**85.68**

4.1   Results of News Quotes Dataset

Table 3 describes results of the models on the ROMIP-2013 news quotes dataset. As it was mentioned before, the participants of the evaluation applied traditional machine learning methods (SVM, Naive Bayes classifier, etc.) and knowledge-based methods with lexicons and rules. The knowledge-based methods achieved the best results. This was explained by thematic diversity of news quotes, when the test collection could contain sentiment words and expressions absent in the training collection. It can be seen in the current evaluation, that the task was difficult even for some models with embeddings (SVM, CNN, LSTM, BiL-STM). Among traditional neural network approaches, BiLSTM obtained the best results.

The use of BERT drastically improves the results. Better results are achieved by conversational RuBERT models. The best configuration is BERT-pair-NLI, when additional *MASK* token is assigned to the whole sentence and the sentence inference task was set.

Table 4. Results on SentiRuEval-2015 Telecom Operators Dataset.

Model	Accuracy	$F_1\ macro$	$F_1^{+-}macro$	$F_1^{+-}micro$
SentiRuEval-2015 [9]	–	–	48.80	53.60
SVM	62.86	58.29	50.27	54.70
CNN	60.80	57.52	49.92	53.23
LSTM	64.46	58.94	52.10	56.03
BiLSTM	65.54	59.35	53.01	56.83
BERT-single	72.48	67.04	58.43	62.53
BERT-pair-QA	74.00	67.83	58.15	62.92
BERT-pair-NLI	74.66	68.24	59.17	64.13
BERT-single (C)	76.55	**69.12**	61.34	66.23
BERT-pair-QA (C)	**76.63**	68.54	**63.47**	**67.51**
BERT-pair-NLI (C)	76.40	68.83	63.14	67.45
Manual	–	–	70.30	70.90

Table 5. Results on SentiRuEval-2015 Banks Dataset.

Model	Accuracy	$F_1\ macro$	$F_1^{+-}macro$	$F_1^{+-}micro$
SentiRuEval-2015 [9]	–	–	36.00	36.60
SVM	49.23	43.39	33.08	36.62
CNN	47.91	42.87	31.62	34.18
LSTM	51.89	44.12	35.85	39.55
BiLSTM	53.21	46.43	36.93	40.18
BERT-single	83.78	74.57	57.82	60.64
BERT-pair-QA	84.24	75.34	56.65	57.41
BERT-pair-NLI	85.14	77.59	60.46	63.15
BERT-single (C)	85.80	78.71	64.90	66.95
BERT-pair-QA (C)	86.28	78.62	62.37	67.27
BERT-pair-NLI (C)	**86.88**	**79.51**	**67.44**	**70.09**

4.2   Results on Twitter Datasets

Tables 4 and 5 describe results of the models on two Twitter datasets of
SentiRuEval-2015. The specific feature of this evaluation was a long 6 months
period of time between downloading the training and test collections. In this
period Ukrainian topics of tweets about telecom operators and banks led to
great differences between the training and test collections.

These differences between collections showed up in very low obtained results
on the bank 2015 dataset [9]. The problem was also complicated for the current
SVM+FastText, CNN, LSTM and BiLSTM models. Only BERT-based methods

Table 6. Results on SentiRuEval-2016 Telecom Operators Dataset.

Model	Accuracy	F_1 macro	F_1^{+-} macro	F_1^{+-} micro
SentiRuEval-2016 [10]	–	–	55.94	65.69
SVM	65.89	55.34	53.13	65.87
CNN	65.28	54.87	52.62	64.40
LSTM	66.71	56.74	56.93	67.18
BiLSTM	67.30	57.11	57.23	67.93
BERT-single	72.85	65.12	60.29	71.70
BERT-pair-QA	74.24	66.34	63.86	73.26
BERT-pair-NLI	74.51	67.48	62.81	73.39
BERT-single (C)	75.20	67.89	64.96	73.91
BERT-pair-QA (C)	75.27	68.11	65.91	**74.22**
BERT-pair-NLI (C)	**75.71**	**68.42**	**66.07**	74.11

could significantly improved the results. Conversational RuBERT in the NLI setting was the best method again.

It is interesting to note that one participant of the SentiRuEval-2015 uploaded manual annotation of the test Telecom dataset and obtained the results described in Table 4 as Manual [9]. It can be seen that the best BERT results are very close to the manual labeling.

Table 7. Results on SentiRuEval-2016 Banks Dataset.

Model	Accuracy	F_1 macro	F_1^{+-} macro	F_1^{+-} micro
SentiRuEval-2016 [10]	–	–	55.17	58.81
SVM	66.46	57.85	51.12	53.74
CNN	67.15	58.43	52.06	54.96
LSTM	70.80	61.17	57.22	59.71
BiLSTM	71.44	61.86	58.40	61.06
BERT-single	81.20	73.21	68.19	69.56
BERT-pair-QA	80.35	72.61	66.61	68.18
BERT-pair-NLI	80.91	72.68	65.62	67.65
BERT-single (C)	80.47	72.59	66.95	69.46
BERT-pair-QA (C)	**82.28**	**74.06**	**69.53**	**71.76**
BERT-pair-NLI (C)	81.28	73.34	65.82	68.03

Tables 6 and 7 describe results of the models on two Twitter datasets of SentiRuEval-2016. In contrast to previous evaluations, baseline results of the 2016 competition (the best results achieved by participants) are better than the SVM+FastText and CNN models. This is due to the fact that the participants applied neural network architectures with embeddings and combined the SVM method with existing Russian sentiment lexicons [2, 10].

5   Analysis of Difficult Examples

The authors of previous Russian sentiment evaluations described examples, which were difficult for most participants of the shared tasks [4, 9, 10]. We gathered these examples and obtained the collection of 21 difficult samples. Now we can compare the performance of the models on this collection.

The difficult examples are translated from Russian and can be subdivided into several groups.

The first type of difficulties concerns the problem of the absence of a sentiment word or word with positive or negative connotations in the training collection, which was a serious problem for previous approaches. From this group one example was again erroneously classified by all current models:

– *"Sberbank imposes credit cards"*. (Ex.1)

The following sentence from this group was successfully processed by all models:

– *"In the capital there was a daring robbery of Sberbank"*. (Ex.2)

The second groups comprises examples with complicated word combinations that include words of different sentiments and/or sentiment operators. From these examples, the following example was problematic for all models:

– *"Secretary of the Presidium of the General Council of United Russia, State Duma Deputy Chairman Sergei Neverov said on Saturday that the party is **not afraid of a split** due to the appearance different ideological platforms in it"*. (Ex.3)

In the above-mentioned sentence there are two negative words and negation, which inverts negative sentiment to positive: *"not afraid of a split"*. But the following example was processed correctly by most BERT-based models:

– *"VTB-24 **reduced losses** in the second quarter"*. (Ex.4)

The third group includes tweets with irony. The following example was differently treated by the models:

– *"Sberbank – the largest network of non-working ATMs in Russia"*. (Ex.5)

The fourth group includes tweets that mention two telecom operators with different sentiment attitudes. In most cases it was difficult for models to distinguish correct sentiment towards each company.

– *"I always said to you that the best operator is Beeline. Megaphone does not respect you"*. (Ex.6 - Beeline, Ex.7 - Megaphone)

Table 8. Analysis of difficult examples. "Acc." means the accuracy of classification on whole collection of 21 difficult examples.

Example	True	SVM	CNN	LSTM	BiLSTM	BS	BPQ	BPN	BS-C	BPQ-C	BPN-C
Ex.1	−1	0	0	0	0	0	0	0	0	0	0
Ex.2	−1	−1	−1	−1	−1	−1	−1	−1	−1	−1	−1
Ex.3	1	−1	−1	−1	−1	−1	−1	−1	−1	−1	−1
Ex.4	1	−1	−1	0	−1	1	−1	1	1	−1	1
Ex.5	−1	−1	−1	−1	−1	0	0	0	0	−1	−1
Ex.6	1	0	0	0	0	−1	−1	−1	−1	−1	−1
Ex.7	−1	0	0	−1	−1	−1	−1	−1	−1	−1	−1
Acc.		0.33	0.24	0.48	0.52	0.48	0.53	0.62	0.62	0.57	**0.71**

Table 8 describes the results of the models on difficult examples. Due to limited space, acronyms of corresponding BERT architectures from previous tables were used. Here $−1, 0, 1$ denote negative, neutral and positive sentiments respectively. Correct predictions are in bold. The last row is share of correct answers for each model. The best results are achieved by BERT-pair-NLI model with pre-trained Conversational RuBERT.

6   Related Work

The most latest and largest Russian sentiment dataset is RuSentiment [14], which contains more than 30000 posts from VKontakte (VK), the most popular social network in Russia. Each post is labeled with one of five classes. The authors evaluated several traditional machine learning methods (logistic regression, linear SVM, Gradient Boosting) and neural networks. The best result (71.7 F_1 measure) was achieved by the neural network with four full-connected layers and FastText embeddings trained on VKontakte posts. In [7] the authors applied to the RuSentiment dataset multilingual BERT and RuBERT, trained on Russian text collections and obtained F_1 measure 87.73 by RuBERT.

Another popular dataset for Russian sentiment analysis is a tweet collection with automatic annotations based on emoticons (RuTweetCorp) [16]. This corpus contains more than 200 thousand Twitter messages posted in 2013–2014 annotated as positive and negative.

In [17], SVM with Word2Vec embeddings were applied the RuTweetCorp dataset. The authors of [21] tested LSTM+CNN and BiGRU models on RuSentiment and RuTweetCorp datasets. Zvonarev and Bilyi [23] compared logistic regression, XGBoost classifier and Convolutional Neural Network on RuTweetCorp and obtained the best results with CNN.

The authors of [11] created the RuSentRel corpus consisted of analytical articles devoted to international relations. The corpus is annotated with sentiment attitudes towards mentioned named entities. Rusnachenko et al. [18] study extraction of sentiment attitudes using CNN and distant supervision approach on the RuSentRel corpus.

7   Conclusion

In this study, we tested standard neural network architectures (CNN, LSTM, BiLSTM) and recently appeared BERT models on previous Russian sentiment evaluation datasets. We applied not only vanilla BERT classification approach, but reformulation of the classification task and question-answering (QA) and natural-language inference (NLI) tasks. We also compared two variants of Russian BERT and showed that for all sentiment tasks in this study the conversational variant of Russian BERT is better.

The best results were mostly achieved by BERT-NLI model. In one of the tasks this model practically achieved the human level of sentiment analysis.

The source code[6] and all sentiment datasets[7] used in this work are publicly available.

Acknowledgments. The reported study was funded by RFBR according to the research project № 20-07-01059.

References

1. Amigó, E., et al.: Overview of RepLab 2013: evaluating online reputation monitoring systems. In: Forner, P., Müller, H., Paredes, R., Rosso, P., Stein, B. (eds.) CLEF 2013. LNCS, vol. 8138, pp. 333–352. Springer, Heidelberg (2013). https://doi.org/10.1007/978-3-642-40802-1_31

2. Arkhipenko, K., Kozlov, I., Trofimovich, J., Skorniakov, K., Gomzin, A., Turdakov, D.: Comparison of neural network architectures for sentiment analysis of Russian tweets. In: Computational Linguistics and Intellectual Technologies: Proceedings of the International Conference Dialogue, pp. 50–59 (2016)

3. Burtsev, M.: DeepPavlov: open-source library for dialogue systems. In: Proceedings of ACL 2018, System Demonstrations, pp. 122–127 (2018)

4. Chetviorkin, I., Loukachevitch, N.: Evaluating sentiment analysis systems in Russian. In: Proceedings of the 4th Biennial International Workshop on Balto-Slavic Natural Language Processing, pp. 12–17 (2013)

5. Cliche, M.: BB twtr at SemEval-2017 task 4: twitter sentiment analysis with CNNs and LSTMs. In: Proceedings of the 11th International Workshop on Semantic Evaluation (SemEval-2017), pp. 573–580 (2017)

6. Devlin, J., Chang, M.W., Lee, K., Toutanova, K.: BERT: Pre-training of Deep Bidirectional Transformers for Language Understanding. arXiv preprint arXiv:1810.04805 (2018)

7. Kuratov, Y., Arkhipov, M.: Adaptation of deep bidirectional multilingual transformers for Russian language (2019)

8. Kuznetsova, E., Loukachevitch, N., Chetviorkin, I.: Testing rules for a sentiment analysis system. In: Proceedings of International Conference Dialog, pp. 71–80 (2013)

---

[6] https://github.com/antongolubev5/Targeted-SA-for-Russian-Datasets.
[7] https://github.com/LAIR-RCC/Russian-Sentiment-Analysis-Evaluation-Datasets.

9. Loukachevitch, N., Rubtsova, Y.: Entity-oriented sentiment analysis of tweets: results and problems. In: Král, P., Matoušek, V. (eds.) TSD 2015. LNCS (LNAI), vol. 9302, pp. 551–559. Springer, Cham (2015). https://doi.org/10.1007/978-3-319-24033-6_62

10. Loukachevitch, N., Rubtsova, Y.: SentiRuEval-2016: overcoming time gap and data sparsity in tweet sentiment analysis. In: Proceedings of International Conference Dialog-2016 (2016)

11. Loukachevitch, N., Rusnachenko, N.: Extracting sentiment attitudes from analytical texts. In: Proceedings of Computational Linguistics and Intellectual Technologies, Papers from the Annual Conference Dialog-2018, pp. 459–468 (2018)

12. Maas, A., Daly, R., Pham, P., Huang, D., Ng, A.Y., Potts, C.: Learning word vectors for sentiment analysis. In: Proceedings of the 49th Annual Meeting of the Association for Computational Linguistics, vol. 1, pp. 142–150 (2011)

13. Nakov, P., Ritter, A., Rosenthal, S., Sebastiani, F., Stoyanov, V.: Semeval-2016 task 4: sentiment analysis in twitter. In: Proceedings of the 10th International Workshop on Semantic Evaluations, SemEval-2016, pp. 502–518 (2016)

14. Rogers, A., Romanov, A., Rumshisky, A., Volkova, S., Gronas, M., Gribov, A.: RuSentiment: an enriched sentiment analysis dataset for social media in Russian. In: Proceedings of the 27th International Conference on Computational Linguistics, pp. 755–763 (2018)

15. Rosenthal, S., Farra, N., Nakov, P.: Semeval-2017 task 4: sentiment analysis in twitter. In: Proceedings of the 11th International Workshop on Semantic Evaluations (SemEval-2017) (2017)

16. Rubtsova, Y.: Constructing a corpus for sentiment classification training. Softw. Syst. **109**, 72–78 (2015)

17. Rubtsova, Y.: Reducing the deterioration of sentiment analysis results due to the time impact. Information **9**(8), 184 (2018)

18. Rusnachenko, N., Loukachevitch, N., Tutubalina, E.: Distant supervision for sentiment attitude extraction. In: Proceedings of the International Conference on Recent Advances in Natural Language Processing (RANLP 2019), pp. 1022–1030 (2019)

19. Socher, R., et al.: Recursive deep models for semantic compositionality over a sentiment treebank. In: Proceedings of the 2013 Conference on Empirical Methods in Natural Language Processing, pp. 1631–1642 (2013)

20. Sun, C., Huang, L., Qiu, X.: Utilizing BERT for aspect-based sentiment analysis via constructing auxiliary sentence. In: Proceedings of the 2019 Conference of the North American Chapter of the Association for Computational Linguistics: Human Language Technologies, vol. 1, pp. 380–385 (2019)

21. Svetlov, K., Platonov, K.: Sentiment analysis of posts and comments in the accounts of Russian politicians on the social network. In: 2019 25th Conference of Open Innovations Association (FRUCT), pp. 299–305. IEEE (2019)

22. Zhang, Y., Wallace, B.: A Sensitivity Analysis of (and Practitioners' Guide to) Convolutional Neural Networks for Sentence Classification. arXiv preprint arXiv:1510.03820, pp. 573–580 (2015)

23. Zvonarev, A., Bilyi, A.: A comparison of machine learning methods of sentiment analysis based on Russian language twitter data. In: The 11th Majorov International Conference on Software Engineering and Computer Systems (2019)

Dataset for Automatic Summarization of Russian News

Ilya Gusev[(⊠)] [iD]

Moscow Institute of Physics and Technology, Moscow, Russia
`ilya.gusev@phystech.edu`

Abstract. Automatic text summarization has been studied in a variety of domains and languages. However, this does not hold for the Russian language. To overcome this issue, we present Gazeta, the first dataset for summarization of Russian news. We describe the properties of this dataset and benchmark several extractive and abstractive models. We demonstrate that the dataset is a valid task for methods of text summarization for Russian. Additionally, we prove the pretrained mBART model to be useful for Russian text summarization.

Keywords: Text summarization · Russian language · Dataset · mBART

1 Introduction

Text summarization is the task of creating a shorter version of a document that captures essential information. Methods of automatic text summarization can be extractive or abstractive.

Extractive methods copy chunks of original documents to form a summary. In this case, the task usually reduces to tagging words or sentences. The resulting summary will be grammatically coherent, especially in the case of sentence copying. However, this is not enough for high-quality summarization as a good summary should paraphrase and generalize an original text.

Recent advances in the field are usually utilizing abstractive models to get better summaries. These models can generate new words that do not exist in original texts. It allows them to compress text in a better way via sentence fusion and paraphrasing.

Before the dominance of sequence-to-sequence models [1], the most common approach was extractive.

The approach's design allows us to use classic machine learning methods [2], various neural network architectures such as RNNs [3,4] or Transformers [5], and pretrained models such as BERT [6,8]. The approach can still be useful on some datasets, but modern abstractive methods outperform extractive ones on CNN/DailyMail dataset since Pointer-Generators [7]. Various pretraining tasks such as MLM (masked language model) and NSP (next sentence prediction) used in BERT [6] or denoising autoencoding used in BART [9] allow models

A. Filchenkov et al. (Eds.): AINL 2020, CCIS 1292, pp. 122–134, 2020.
https://doi.org/10.1007/978-3-030-59082-6_9

to incorporate rich language knowledge to understand original documents and generate grammatically correct and reasonable summaries.

In recent years, many novel text summarization datasets have been revealed. XSum [11] focuses on very abstractive summaries; Newsroom [12] has more than a million pairs; Multi-News [13] reintroduces multi-document summarization. However, datasets for any language other than English are still scarce. For Russian, there are only headline generation datasets such as RIA corpus [14]. The main aim of this paper is to fix this situation by presenting a Russian summarization dataset and evaluating some of the existing methods on it.

Moreover, we adapted the mBART [10] model initially used for machine translation to the summarization task. The BART [9] model was successfully used for text summarization on English datasets, so it is natural for mBART to handle the same task for all trained languages.

We believe that text summarization is a vital task for many news agencies and news aggregators. It is hard for humans to compose a good summary, so automation in this area will be useful for news editors and readers. Furthermore, text summarization is one of the benchmarks for general natural language understanding models.

Our contributions are as follows: we introduce the first Russian summarization dataset in the news domain.[1] We benchmark extractive and abstractive methods on this dataset to inspire further work in the area. Finally, we adopt the mBART model to summarize Russian texts, and it achieves the best results of all benchmarked models.[2]

2   Data

2.1   Source

There are several requirements for a data source. First, we wanted news summaries as most of the datasets in English are in this domain. Second, these summaries should be human-generated. Third, no legal issues should exist with data and its publishing. The last requirement was hard to fulfill as many news agencies have explicit restrictions for publishing their data and tend not to reply to any letters.

Gazeta.ru was one of the agencies with explicit permission on their website to use their data for non-commercial purposes. Moreover, they have summaries for many of their articles.

There are also requirements for content of summaries. We do not want summaries to be fully extractive, as it would be a much easier task, and consequently, it would not be a good benchmark for abstractive models.

We collected texts, dates, URLs, titles, and summaries of all articles from the website's foundation to March 2020. We parsed summaries as the content of a "meta" tag with "description" property. A small percentage of all articles had a summary.

---

[1] https://github.com/IlyaGusev/gazeta.
[2] https://github.com/IlyaGusev/summarus.

2.2 Cleaning

After the scraping, we did cleaning. We removed summaries with more than 85 words and less than 15 words, texts with more than 1500 words, pairs with less than 30% unigram intersection, and more than 92% unigram intersection. The examples outside these borders contained either fully extractive summaries or not summaries at all. Moreover, we removed all data earlier than the 1st of June 2010 because the meta tag texts were not news summaries. The complete code of a cleaning phase is available online with a raw version of the dataset.

2.3 Statistics

The resulting dataset consists of 63435 text-summary pairs. To form training, validation, and test datasets, these pairs were sorted by time. We define the first 52400 pairs as the training dataset, the proceeding 5265 pairs as the validation dataset, and the remaining 5770 pairs as the test dataset. It is still essential to randomly shuffle the training dataset before training any models to reduce time bias even more.

Statistics of the dataset can be seen in Table 1. Summaries of the training part of the dataset are shorter on average than summaries of validation and test parts. We also provide statistics on lemmatized texts and summaries. We compute normal forms of words using the pymorphy2 [28][3] package. Numbers in the "Common UL" row show size of an intersection between lemmas' vocabularies of texts and summaries. These numbers are almost similar to numbers in the "Unique lemmas" row of summaries' columns. It means that almost all lemmas of the summaries are presented in original texts.

Table 1. Dataset statistics after lowercasing

	Train		Validation		Test	
	Text	Summary	Text	Summary	Text	Summary
Dates	01.06.10–31.05.19		01.06.19–30.09.19		01.10.19–23.03.20	
Pairs	52 400		5265		5770	
Unique words: UW	611 829	148 073	167 612	42 104	175 369	44 169
Unique lemmas: UL	282 867	63 351	70 210	19 698	75 214	20 637
Common UL	60 992		19 138		20 098	
Min words	28	15	191	18	357	18
Max words	1500	85	1500	85	1498	85
Avg words	766.5	48.8	772.4	54.5	750.3	53.2
Avg sentences	37.2	2.7	38.5	3.0	37.0	2.9
Avg UW	419.1	41.3	424.2	46.0	415.7	45.1
Avg UL	350.0	40.2	352.5	44.6	345.4	43.9

---

[3] https://github.com/kmike/pymorphy2.

We depict the distribution of tokens counts in texts in Fig. 1, and the distribution of tokens counts in summaries is in Fig. 2. The training dataset has a smoother distribution of text lengths in comparison with validation and test datasets. It also has an almost symmetrical distribution of summaries' lengths, while validation and test distributions are skewed.

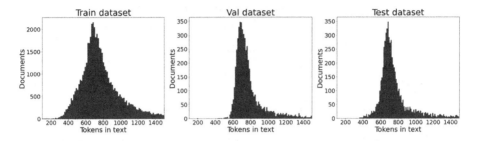

Fig. 1. Documents distribution by count of tokens in a text

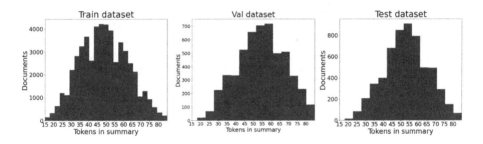

Fig. 2. Documents distribution by count of tokens in a summary

To evaluate the dataset's bias towards extractive or abstractive methods, we measured the percentage of novel n-grams in summaries. Results are presented in Table 2 and show that more than 65% of summaries' bi-grams do not exist in original texts. This number decreases to 58% if we consider different word forms and calculate it on lemmatized bi-grams. Although we can not directly compare these numbers with CNN/DailyMail or any other English dataset as this statistic is heavily language-dependent, we should state that it is 53% for CNN/DailyMail and 83% for XSum. From this, we can conclude that the bias towards extractive methods can exist.

Another way to evaluate the abstractiveness is by calculating metrics of oracle summaries (the term is defined in Sect. 3.2). To evaluate all benchmark models, we used ROUGE [22] metrics. For CNN/DailyMail oracle summaries score 31.2 ROUGE-2-F [8], and for our dataset, it is 22.7 ROUGE-2-F.

Table 2. Average % of novel n-grams

	Train	Val	Test
Uni-grams	34.2	30.5	30.6
Lemmatized uni-grams	21.4	17.8	17.6
Bi-grams	68.6	65.0	65.5
Lemmatized bi-grams	61.4	58.0	58.5
Tri-grams	84.5	81.5	81.9

2.4 BPE

We extensively utilized byte-pair encoding (BPE) tokenization in most of the described models. For Russian, the models that use BPE tokenization performs better than those that use word tokenization as it enables the use of rich morphology and decreases the number of unknown tokens. The encoding was trained on the training dataset using the sentencepiece [25] library.

2.5 Lowercasing

We lower-cased all texts and summaries in most of our experiments. It is a controversial decision. On the one hand, we reduced vocabulary size and focused on the essential properties of models, but on the other hand, we lost important information for a model to receive. Moreover, if we speak about our summarization system's possible end-users, it is better to generate summaries in the original case.

We provide a non-lower-cased version of the dataset as the main version for possible future research.

3 Benchmark Methods

We used several groups of methods. TextRank [15] and LexRank [16] are fully unsupervised extractive summarization methods. Summarunner [4] is a supervised extractive method. PG [7], CopyNet [20], mBART [10] are abstractive summarization methods.

3.1 Unsupervised Methods

This group of methods does not have any access to reference summaries and utilizes only original texts. All of the considered methods in this group extract whole sentences from a text, not separated words.

TextRank. TextRank [15] is a classic graph-based method for unsupervised text summarization. It splits a text into sentences, calculates a similarity matrix for every distinct pair of them, and applies the PageRank algorithm to obtain final scores for every sentence. After that, it takes the best sentences by the score as a predicted summary. We used TextRank implementation from the summa [17][4] library. It defines sentence similarity as a function of a count of common words between sentences and lengths of both sentences.

LexRank. Continuous LexRank [16] can be seen as a modification of the TextRank that utilizes TF-IDF of words to compute sentence similarity as IDF modified cosine similarity. A continuous version uses an original similarity matrix, and a base version performs binary discretization of this matrix by the threshold. We used LexRank implementation from lexrank Python package.[5]

LSA. Latent semantic analysis can be used for text summarization [21]. It constructs a matrix of terms by sentences with term frequencies, applies singular value decomposition to it, and searches right singular vectors' maximum values. The search represents finding the best sentence describing the k'th topic. We evaluated this method with sumy library.[6]

3.2 Supervised Extractive Methods

Methods in this group have access to reference summaries, and the task for them is seen as sentences' binary classification. For every sentence in an original text, the algorithm must decide whether to include it in the predicted summary.

To perform the reduction to this task, we first need to find subsets of original sentences that are most similar to reference summaries. To find these so-called "oracle" summaries, we used a greedy algorithm similar to SummaRuNNer paper [4] and BertSumExt paper [8]. The algorithm generates a summary consisting of multiple sentences which maximize the ROUGE-2 score against a reference summary.

SummaRuNNer. SummaRuNNer [4] is one of the simplest and yet effective neural approaches to extractive summarization. It uses 2-layer hierarchical RNN and positional embeddings to choose a binary label for every sentence. We used our implementation on top of the AllenNLP [19][7] framework along with Pointer-Generator [7] implementation.

---

[4] https://github.com/summanlp/textrank.
[5] https://github.com/crabcamp/lexrank.
[6] https://github.com/miso-belica/sumy.
[7] https://github.com/allenai/allennlp.

3.3   Abstractive Methods

All of the tested models in this group are based on a sequence-to-sequence framework. Pointer-generator and CopyNet were trained only on our training dataset, and mBART was pretrained on texts of 25 languages extracted from the Common Crawl. We performed no additional pretraining, though it is possible to utilize Russian headline generation datasets here.

Pointer-Generator. Pointer-generator [7] is a modification of a sequence-to-sequence RNN model with attention [18]. The generation phase samples words not only from the vocabulary but from the source text based on attention distribution. Furthermore, the second modification, the coverage mechanism, prevents the model from attending to the same places many times to handle repetition in summaries.

CopyNet. CopyNet [20] is another variation of sequence-to-sequence RNN model with attention with slightly different copying mechanism. We used the stock implementation from AllenNLP [19].

MBART for Summarization. BART [9] and mBART [10] are sequence-to-sequence Transformer models with autoregressive decoder trained on the denoising autoencoding task. Unlike the preceding pretrained models like BERT, they focus on text generation even in the pretraining phase.

mBART was pretrained on the monolingual corpora for 25 languages, including Russian. In the original paper, it was successfully used for machine translation. BART was used for text summarization, so it is natural to try a pretrained mBART model for Russian summarization.

We used training and prediction scripts from fairseq [27].[8] However, it is possible to convert the model for using it within HuggingFace's Transformers.[9] We had to truncate input for every text to 600 tokens to fit the model in GPU memory. We also used <unk> token instead of language codes to condition mBART.

4   Results

4.1   Automatic Evaluation

We measured the quality of summarization with three sets of automatic metrics: ROUGE [22], BLEU [23], METEOR [24]. All of them are used in various text generation tasks and are based on the overlaps of N-grams. ROUGE and METEOR are prevalent in text summarization research, and BLEU is a primary automatic metric in machine translation. BLUE is a precision-based metric and

---

[8] https://github.com/pytorch/fairseq.
[9] https://github.com/huggingface/transformers.

does not take recall into account, while ROUGE uses both recall and precision-based metrics in a balanced way, and METEOR weight for the recall part is higher than weight for the precision part.

All three sets of metrics are not perfect as we only have only one version of a reference summary for each text, while it is possible to generate many correct summaries for a given text. Some of these summaries can even have zero n-gram overlap with reference ones.

We lower-cased and tokenized reference and predicted summaries with Razdel tokenizer to unify the methodology across all models. We suggest to all further researchers to use the same evaluation script.

Table 3. Automatic scores for all models on the test set

	ROUGE			BLEU	Meteor
	1	2	L		
Lead-1	27.6	12.9	20.2	19.9	18.6
Lead-2	30.6	13.7	25.6	43.1	23.7
Lead-3	31.0	13.4	26.3	44.2	26.0
Greedy Oracle	44.3	22.7	39.4	53.8	35.5
TextRank	21.4	6.3	16.4	28.6	17.5
LexRank	23.7	7.8	19.9	37.7	18.1
LSA	19.3	5.0	15.0	30.7	15.2
SummaRuNNer	31.6	13.7	27.1	46.3	**26.0**
CopyNet	28.7	12.3	23.6	37.2	21.0
PG small	29.4	12.7	24.6	38.8	21.2
PG words	29.4	12.6	24.4	35.9	20.9
PG big	29.6	12.8	24.6	39.0	21.5
PG small + coverage	30.2	12.9	26.0	42.8	22.7
Finetuned mBART	**32.1**	**14.2**	**27.9**	**50.1**	25.7

We provide all the results in Table 3. Lead-1, lead-2, and lead-3 are the most basic baselines, where we choose the first, the first two, or the first three sentences of every text as our summary. Lead-3 is a strong baseline, as it was in CNN/DailyMail dataset [7]. The oracle summarization is an upper bound for extractive methods.

Unsupervised methods give summaries that are very dissimilar to the original ones. LexRank is the best of unsupervised methods in our experiments.

The SummRuNNer model has the best METEOR score and high BLEU and ROUGE scores. In Fig. 3, SummaRuNNer has a bias towards the sentences at the beginning of the text compared to the oracle summaries. In contrast, LexRank sentence positions are almost uniformly distributed except for the first sentence.

It seems that more complex extractive models should perform better on this dataset, but unfortunately, we did not have time to prove it.

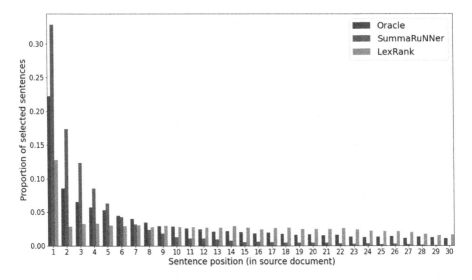

Fig. 3. Proportion of extracted sentences according to their position in the original document.

To evaluate an abstractiveness of the model, we used extraction and plagiarism scores [26]. The plagiarism score is a normalized length of the longest common sequence between a text and a summary. The extraction score is a more sophisticated metric. It computes normalized lengths of all long non-overlapping common sequences between a text and a summary and ensures that the sum of these normalized lengths is between 0 and 1.

As for abstractive models, mBART has the best result among all the models in terms of ROUGE and BLEU. However, Fig. 4 shows that it has fewer novel n-grams than Pointer-Generator with coverage. Consequently, it has worser extraction and plagiarism scores [26] (Table 4).

Table 4. Extraction scores on the test set

	Extraction score	Plagiarism score
Reference	0.031	0.124
PG small + coverage	0.325	0.501
Finetuned mBART	0.332	0.502
SummaRuNNer	0.513	0.662

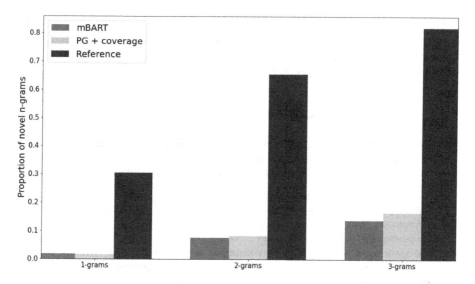

Fig. 4. Proportion of novel n-grams in model generated summaries on the test set

4.2 Human Evaluation

We also did side-by-side annotation of mBART and human summaries with Yandex.Toloka,[10] a Russian crowdsourcing platform. We sampled 1000 text and summary pairs from the test dataset and generated a new summary for every text. We showed a title, a text, and two possible summaries for every example. Nine people annotated every example. We asked them which summary is better and provided them three options: left summary wins, draw, right summary wins. The side of the human summary was random. Annotators were required to pass training, exam, and their work was continuously evaluated through the control pairs ("honeypots").

Table 5 shows the results of the annotation. There were no full draws, so we exclude them from the table. mBART wins in more than 73% cases. We cannot just conclude that it performs on a superhuman level from these results. We did not ask our annotators to evaluate the abstractiveness of the summaries in any way. Reference summaries are usually too provocative and subjective, while mBART generates highly extractive summaries without any errors and with many essential details, and annotators tend to like it. The annotation task should be changed to evaluate the abstractiveness of the model. Even so, that is an excellent result for mBART.

Table 5. Human side-by-side evaluation

Votes for winner	Reference wins	mBART wins
Majority	265	735
9/9	7	47
8/9	18	106
7/9	30	185
6/9	54	200
5/9	123	180
4/9	32	17
3/9	1	0

---

[10] https://toloka.yandex.ru/.

Table 6 shows examples of mBART losses against reference summaries. In the first example, there is an unnamed entity in the first sentence, "by them" ("*ими*"). In the second example, the factual error and repetition exist. In the last example, the last sentence is not cohesive.

Table 6. mBART summaries that lost 9/9

разработанный ими метод идентификации способен выделить специфические для конкретного человека белки из пряди волос длиной всего сантиметра . это позволит с высокой степенью точности идентифицировать людей и без выделения днк .

президент россии владимир путин на встрече с ветеранами и представителями общественных патриотических объединений заявил , что каждый год единовременные выплаты ко дню победы составляют по 10 тыс . рублей ветеранам и по 5 тыс . рублей труженикам тыла . по 50 тыс . рублей также будет выплачено труженикам тыла . ранее в послании федеральному собранию президент также подчеркнул важность предстоящего юбилея вов .

самый одинокий актер голливуда , наконец , официально нашел пару . киану ривз , который многие годы предпочитал не распространяться о своей личной жизни и после давней трагедии решил не иметь детей , пришел на светское мероприятие с 46-летней художницей из лос-анджелеса александрой грант , чем вызвал ажиотаж у журналистов .', 'на арт-ивенте lacma art + film gala , прошедшем при поддержке gucci , актер киану ривз завел девушку — впервые за последние 20 лет . по словам артиста , в этом кругу редко вращается и ривз , несколько лет вызывающий сочувствие пользователей соцсетей фотографиями с празднований своего дня рождения .

5   Conclusion

We present the first corpus for text summarization in the Russian language. We demonstrate that most of the text summarization methods work well for Russian without any special modifications. Moreover, mBART performs exceptionally well even if it was not initially designed for text summarization in the Russian language.

We wanted to extend the dataset using data from other sources, but there were significant legal issues in most cases, as most of the sources explicitly forbid any publishing of their data even in non-commercial purposes.

In future work, we will pre-train BART ourselves on standard Russian text collections and open news datasets. Furthermore, we will try the headline generation as a pretraining task for this dataset. We believe it will increase the performance of the models.

References

1. Sutskever, I., Vinyals, I., Le, Q.: Sequence to sequence learning with neural networks. In: Proceedings of the 27th International Conference on Neural Information Processing Systems, vol. 2, pp. 3104–3112. MIT Press, Cambridge (2014)
2. Wong, K., Wu, M., Li W.: Extractive summarization using supervised and semi-supervised learning. In: Proceedings of the 22nd International Conference on Computational Linguistics, pp. 985–992. Coling 2008 Organizing Committee (2008)
3. Hochreiter, S., Schmidhuber, J.: Long short-term memory. Neural Comput. **9**(8), 1735–1780 (1997)
4. Nallapati, R., Zhai, F., Zhou B.: SummaRuNNer: a recurrent neural network based sequence model for extractive summarization of documents. In: Proceedings of the Thirty-First AAAI Conference on Artificial Intelligence, pp. 3075–3081 (2017)
5. Vaswani, A., et al.: Attention is all you need. In: Advances in Neural Information Processing Systems, pp. 5998–6008 (2017)
6. Devlin, J., Chang, M., Lee, K., Toutanova, K.: BERT: pre-training of deep bidirectional transformers for language understanding. In: Proceedings of the 2019 Conference of the North American Chapter of the Association for Computational Linguistics: Human Language Technologies, Minneapolis, Minnesota, vol. 1, pp. 4171–4186 (2019)
7. See, A., Liu, P., Manning, C.: Get to the point: summarization with pointer-generator networks. In: Proceedings of the 55th Annual Meeting of the Association for Computational Linguistics, vol. 1, pp. 1073–1083. Association for Computational Linguistics, Vancouver (2017)
8. Liu, Y., Lapata, M.: Text summarization with pretrained encoders. In: Proceedings of the 2019 Conference on Empirical Methods in Natural Language Processing and the 9th International Joint Conference on Natural Language Processing (EMNLP-IJCNLP), pp. 3730–3740. Association for Computational Linguistics, Hong Kong (2019)
9. Lewis, M., et al.: BART: denoising sequence-to-sequence pre-training for natural language generation, translation, and comprehension. In: Proceedings of the 2019 Conference on Empirical Methods in Natural Language Processing and the 9th International Joint Conference on Natural Language Processing (EMNLP-IJCNLP), pp. 4003–4015. Association for Computational Linguistics, Hong Kong (2019)
10. Liu, Y., et al.: Multilingual Denoising Pre-training for Neural Machine Translation. arXiv preprint arXiv:2001.08210 (2020)
11. Narayan, S., Cohen, S., Lapata, M.: Don't give me the details, just the summary! topic-aware convolutional neural networks for extreme summarization. In: Proceedings of the 2018 Conference on Empirical Methods in Natural Language Processing, Brussels (2018)
12. Grusky, M., Naaman, M., Artzi, Y.: NEWSROOM: a dataset of 1.3 million summaries with diverse extractive strategies. In: Proceedings of the 2018 Conference of the American Chapter of the Association for Computational Linguistics: Human Language Technologies. Association for Computational Linguistics, New Orleans (2018)
13. Fabbri, A., Li, I., She, T., Li, S., Radev, D.: Multi-news: a large-scale multi-document summarization dataset and abstractive hierarchical mode. In: Proceedings of the 57th Annual Meeting of the Association for Computational Linguistics, pp. 1074–1084. Association for Computational Linguistics, Florence (2019)

14. Azzopardi, L., Stein, B., Fuhr, N., Mayr, P., Hauff, C., Hiemstra, D. (eds.): ECIR 2019. LNCS, vol. 11437. Springer, Cham (2019). https://doi.org/10.1007/978-3-030-15712-8

15. Mihalcea, R., Tarau, P.: TextRank: bringing order into text. In: Proceedings of the 2004 Conference on Empirical Methods in Natural Language Processing, pp. 404–411. Association for Computational Linguistics, Barcelona (2004)

16. Erkan, G., Radev, D.: LexRank: graph-based lexical centrality as salience in text summarization. J. Artif. Intell. Res. **22**(1), 457–479 (2004). AI Access Foundation

17. Barrios, F., López, F., Argerich, L., Wachenchauzer, R.: Variations of the Similarity Function of TextRank for Automated Summarization. arXiv preprint arXiv:1602.03606 (2016)

18. Bahdanau, D., Cho, K., Bengio, Y.: Neural machine translation by jointly learning to align and translate. In: International Conference on Learning Representations (2015)

19. Gardner, M., et al.: AllenNLP: A Deep Semantic Natural Language Processing Platform. arXiv preprint arXiv:1803.07640 (2018)

20. Gu, J., Lu, Z., Li, H., Li, V.: Incorporating copying mechanism in sequence-to-sequence learning. In: Proceedings of the 54th Annual Meeting of the Association for Computational Linguistics, vol. 1, pp. 1631–1640. Association for Computational Linguistics (2016)

21. Gong, Y., Liu, X.: Generic text summarization using relevance measure and latent semantic analysis. In: Proceedings of the 24th Annual International ACM SIGIR Conference on Research and Development in Information Retrieval, pp. 19–25 (2001)

22. Lin, C.: ROUGE: a package for automatic evaluation of summaries. In: Text Summarization Branches Out, pp. 74–81. Association for Computational Linguistics, Barcelona (2004)

23. Papineni, K., Roukos, S., Ward, T., Zhu, W. J.: BLEU: a method for automatic evaluation of machine translation. In: 40th Annual Meeting of the Association for Computational Linguistics, pp. 311–318 (2002)

24. Denkowski, M., Lavie, A.: Meteor universal: language specific translation evaluation for any target language. In: Proceedings of the EACL 2014 Workshop on Statistical Machine Translation (2014)

25. Kudo, T., Richardson, J.: SentencePiece: a simple and language independent subword tokenizer and detokenizer for neural text processing. In: Proceedings of the 2018 Conference on Empirical Methods in Natural Language Processing: System Demonstrations, pp. 66–71 (2018)

26. Cibils, A., Musat, C., Hossmann, A., Baeriswyl, M.: Diverse beam search for increased novelty in abstractive summarization. arXiv preprint arXiv:1802.01457 (2018)

27. Ott, M., et al.: fairseq: a fast, extensible toolkit for sequence modeling. In: Proceedings of NAACL-HLT 2019: Demonstrations (2019)

28. Korobov, M.: Morphological analyzer and generator for Russian and Ukrainian languages. In: Analysis of Images, Social Networks and Texts, pp 320–332 (2015)

Dataset for Evaluation of Mathematical Reasoning Abilities in Russian

Mikhail Nefedov[✉][iD]

Higher School of Economics, Moscow, Russia
manefedov26@gmail.com

Abstract. We present a Russian version of DeepMind Mathematics Dataset. The original dataset is synthetically generated using inference rules and a set of linguistic templates. We translate the linguistic templates to Russian leaving the inference part without changes. So as a result we get a mathematically parallel dataset where the same mathematical problems are explored but in another language. We reproduce the experiment from the original paper to check whether the performance of a Transformer model is impacted by the differences of the languages in which math problems are expressed. Though our contribution is small compared to the original work, we think it is valuable given the fact that languages other than English (and Russian in particular) are underrepresented.

1 Introduction

Research in natural language processing relies heavily on the availability of high quality datasets. They are used as benchmarks for testing novel architectures and pretrained models.

New datasets are constantly being created to explore more and more complex and nuanced problems. However, most of the datasets are in English. According to several analyses [1,7,8], most of the papers published at ACL only deal with English. And major languages like Russian are severely underrepresented. That means that the generalization capabilities of the developed models are underexplored. Models that work with English may not generalize well to other languages.

In this paper we try to make a contribution towards reducing this kind of bias. We adapt the Mathematics Datasets, a large synthetic dataset of mathematical word problems, recently published by DeepMind [13] from English to Russian. Although our contribution is small compared to the original work we think our version of the dataset can have a positive impact. Synthetic English datasets were previously used for evaluation of the ability to generalize in neural networks, for probing of embeddings, for injecting a skill into a pre-trained language model [3].

2 Related Works

As far as the English language is concerned, there are already quite a variety of datasets that aim at testing the ability to reason numerically.

© Springer Nature Switzerland AG 2020
A. Filchenkov et al. (Eds.): AINL 2020, CCIS 1292, pp. 135–144, 2020.
https://doi.org/10.1007/978-3-030-59082-6_10

A number of them are collections of mathematical word problems that were used for teaching. Here are several examples:

- AI2[1] dataset consists of 395 single-step or multi-step arithmetic word problems for the third, fourth, and fifth graders. It involves problems that can be solved with only addition and subtraction.
- SingleEq [5] dataset consists of 508 problems of grade-school algebra. The problems may involve multiple math operations including multiplication, division, subtraction, and addition over non-negative rational numbers and one variable.
- IL [12] dataset contains 562 single-step word problems with only one operator, including addition, subtraction, multiplication, and division.
- CommonCore [11] dataset consists of 600 multi-step word problems of six different types without irrelevant quantities.
- MAWPS [6] dataset contains 3320 word problems but partially overlaps with other existing collections.

Math word problem solving was one of the shared tasks in SemEval2019 [4]. The organizers collected a dateset consisting of 2778 training questions and 1082 test questions.

All the above mentioned datasets are quite small because it is very hard to collect them (the organizers of the SemEval competition mention this in their paper). In a recent paper [13], DeepMind suggested synthetically generating a dataset from a short set of templates and released the largest dataset of mathematical problems. Their work differs from the works mentioned above in that their focus is mathematical reasoning rather than linguistic comprehension, they cover more areas of mathematics but with less variation in problem specification. They used the resulting dataset to evaluate the ability of deep neural networks to generalize.

In [2] numerical reasoning is tested in a more broad context of reading comprehension on the DROP dataset. The dataset consists of 100 000 questions that were written by crowdworkers. The questions refer to textual paragraphs from Wikipedia articles and require discrete reasoning to answer. The authors show that the performance of the models that almost solve SQUAD [9] drops dramatically on DROP. The authors proposed a novel neural architecture NAQANet which is similar to standard reading comprehension architectures but with special layers added to handle a limited number of mathematical operations. A similar approach is used in [15]. The authors proposed a special layer (Neural Arithmetic Logic Unit) to improve the ability of neural networks to learn basic mathematical operations. Even though these approaches show their effectiveness, they do not generalize to other tasks and others types of mathematical operations. [3] suggested using a synthetic dataset to inject the ability to reason numerically into a pre-trained language models (e.g. BERT) without significant changes to the model architecture.

---

[1] http://ai2-website.s3.amazonaws.com/data/arithmeticquestions.pdf.

Synthetic mathematical datasets can also be used to probe embeddings for numeracy [17]. The authors find that popular encoders capture numeracy surprisingly well, given that the numbers are treated as ordinary tokens. However, those models fail to extrapolate beyond what they saw in training. [13] come to the same conclusion while testing Transformer and LSTM models on their dataset.

[10] present EQUATE, an evaluation framework to estimate the ability of models to reason quantitatively in textual entailment. They benchmark the performance of 9 published natural language inference models on 5 test sets, and find that on average, state-of-the-art methods do not achieve an absolute improvement over a majority-class baseline, suggesting that they do not implicitly learn to reason with quantities.

As for the other languages, the field is very sparse. The only dataset we could find is [18] of Chinese math word problems for elementary school students. It consists of 23, 161 math problems with 2, 187 templates. [14] investigates math word problem solving in Arabic. The authors use translated AI2 dataset to evaluate their system.

3   Dataset Translation

The original dataset consists of 56 types of mathematical problems covering the following areas: arithmetic, algebra, calculus, polynomials, measurement, probability, comparison, number theory. The content was based on a national school mathematics curriculum (up to age 16), excluding geometry.

For every problem type there are several similar templates with slots to be filled with a number, an expression, a sequence etc. For example, these templates are used to generate problems on dividing two numbers:

Divide {p} by {q}.

{p} divided by {q}?

What is {p} divided by {q}?

Calculate {p} divided by {q}

where p and q are variable slots that are filled with randomly generated numbers.

So in order to create a Russian version of the dataset we manually translated these templates. In some cases it was possible to make a word by word translation:

Divide {p} by {q}.

Разделите {p} на {q}.

However, it was not always the case and analogous or descriptive expressions needed to be used:

What is the product {p} and {q}?

Каков результат произведения {p} и {q}?

In this example 'product' is translated as 'a result of multiplication' because there is no such term in Russian.

In some cases we ended up with two sentences in translation instead of one:

> What is the cube root of value
> to the nearest integer?
> Чему равен корень кубический {value}?
> Ответ округлите до целого числа.

The problem is divided into two parts: 'What is the cube root of n?' and 'Round the answer to the nearest integer'. Using both in one sentence is possible, but that sentence would be overcomplicated (due to participle clause) and would simply sound awkward.

In general we did not attempt to get exactly the same set of templates for every problem and strove for authenticity. So the sizes of the template lists do not always match. Where it was necessary, the templates were omitted. Where it was possible, more analogous templates were added.

Some of the types were a bit more complicated and we needed to add more code to the generation process. The complicated types are measurement conversions and number rounding. In measurement problems every unit such as length, time or weight is directly modified by the associated number. In English the rule is simply add '-s' when the number is greater than 1. In Russian, however, the rule is: numbers ending with 1 (but not 11) require singular in all cases; numbers ending with 2, 3, 4 require plural form except for the nominal and the accusative cases where singular form is used; other numbers require plural for all cases.

In order to get this right we made case specific templates, added number specific ending rules and extended the lists of unit words with all possible forms. The template for conversion look something like this:

> How many {target_name} are there in
> {base_value} {base_name}?
> Сколько {target_name_genitive}
> в {base_value} {base_name_instr}?

In the rounding problems in addition to the above mentioned complications we also have grammatical gender issues. Words for a ten, a million and a billion are masculine, whereas words for a hundred and thousand are feminine. Therefore, we need to write specific rules to select correct form of 'nearest' for a following template:

> Round {input} to the nearest
> {tenth, hundreds, etc}.
> Округлите {input} до (ближайшего, ближайшей)
> {десятка, сотни}.

In the generation process some numbers are randomly transformed to words (i.e. one instead of 1) to make the problems more diverse. Representing numbers with words in Russian required some more changes to the case-gender-number form selection. For example,

> How many grams are there in
> two thousand three hundred
> and twenty-one kilograms?
> Сколько грамм в двух
> тысячах трёхстах двадцати
> одном килограмме?

Here the words for 'two thousand three hundred and twenty-one' are in the instrumental case. Also note that the word for kilograms is singular because the number ends with 1.

In the example below the words for 'two thousand three hundred and twenty-two' are in the accusative case. The word for kilograms is again in the singular form. As we stated earlier, if a number ends with 2, the following word requires singular in the nominal and the accusative cases:

> Convert two thousand
> three hundred and twenty-two
> kilograms to grams?
> Переведите две тысячи триста
> двадцать два килограмма в граммы.

Another change we had to make is time notation. In Russian there is no equivalent of pm-am for describing time and 12-h format is used only informally. So we used 24-h format in the time templates instead. For example:

> What is 130 minutes before 8:16 PM?
> Сейчас 20:16. Сколько времени было
> 130 минут назад?

What we did not have to change is the math notation. In Russian Latin or Greek symbols are used in math expressions as well:

> Let $t(k) = 16 - 7*k$. Find $t(8)$.
> Пусть $t(k) = 16 - 7*k$. Найдите $t(8)$.

Table 1 summarizes how many templates are there in the original dataset and in our translated version.

All the changes we made did not affect the generation process itself.

Table 1. Number of templates in the original dataset and in the translated version (per module and overall)

Module	Russian templates	English templates
Algebra	18	17
Arithmetic	50	51
Calculus	10	10
Comparison	35	33
Measurement	11	11
Numbers	26	29
Polynomials	10	14
Probability	6	6
Overall	166	171

4   Model Training

We trained a Transformer [16] on both versions of the dataset (English and Russian) in order to see if the language changes introduced significantly affect the performance. The authors of the original dataset also trained a Transformer model and reported the results in the paper [13]. We did not try to compare with their numbers but trained a new model. Number of training steps is lower in our experiments (we didn't have as many computational resources), so our metrics are lower on the English dataset.

We generated two datasets (English and Russian) of almost the same size (200000 examples per training module and 1000 per testing module; the final number of examples can vary a little due to randomness in generation process and maximum length filter). We extended the limit on the question length up to 300 (180 in the original dataset) because Russian sentences are generally longer. Such limit would penalize longer expressions in Russian version. We split the questions and answers on a character level. The vocabulary for Russian model is bigger (88 versus 57 for the question and 54 versus 51 for the answer). We used a Base-Transformer [16] parameters and trained both models for 5 epochs (batch size - 95) using Adam optimizer with custom learning schedule as in [16] with 100000 warm-up steps.

The test sets consist of two parts: the interpolation test set and the extrapolation test set. The interpolation test set contains problems similar to problems in the training set. The extrapolation test set contains problems that are somehow different from the training problems (e.g. they are longer, have more symbols, have larger numbers etc.). The purpose of extrapolation test set is to measure the ability of a model to generalize. Tables 2 and 3 show results for both models on the interpolation and the extrapolation parts of the testing dataset.

The results indicate that there is no major differences in Transformer performance with respect to math word problems in English and in Russian. The

Table 2. Accuracy on interpolation

Module	Accuracy (Rus)	Accuracy (Eng)
Algebra linear 1d	0.69	0.614
Algebra linear 1d composed	0.583	0.566
Algebra linear 2d	0.424	0.309
Algebra linear 2d composed	0.546	0.391
Algebra polynomial roots	0.267	0.267
Algebra polynomial roots composed	0.451	0.513
Algebra sequence next term	0.605	0.564
Algebra sequence nth term	0.402	0.411
Arithmetic add or sub	0.994	0.99
Arithmetic add or sub in base	0.952	0.952
Arithmetic add sub multiple	0.837	0.849
Arithmetic div	0.666	0.665
Arithmetic mixed	0.224	0.226
Arithmetic mul	0.492	0.531
Arithmetic mul div multiple	0.479	0.491
Arithmetic nearest integer root	0.653	0.66
Arithmetic simplify surd	0.018	0.029
Calculus differentiate	0.924	0.939
Calculus differentiate composed	0.728	0.577
Comparison closest	0.955	0.965
Comparison closest composed	0.697	0.726
Comparison kth biggest	0.949	0.972
Comparison kth biggest composed	0.731	0.738
Comparison pair	0.792	0.79
Comparison pair composed	0.643	0.655
Comparison sort	0.942	0.945
Comparison sort composed	0.549	0.587
Measurement conversion	0.85	0.84
Measurement time	0.969	0.98
Numbers base conversion	0.0	0.001
Numbers div remainder	0.186	0.228
Numbers div remainder composed	0.214	0.218
Numbers gcd	0.419	0.41
Numbers gcd composed	0.569	0.555
Numbers is factor	0.724	0.72
Numbers is factor composed	0.62	0.595
Numbers is prime	0.567	0.551
Numbers is prime composed	0.566	0.566
Numbers lcm	0.449	0.392

(*continued*)

Table 2. (*continued*)

Module	Accuracy (Rus)	Accuracy (Eng)
Numbers lcm composed	0.312	0.335
Numbers list prime factors	0.044	0.06
Numbers list prime factors composed	0.114	0.101
Numbers place value	1.0	1.0
Numbers place value composed	0.436	0.414
Numbers round number	1.0	0.999
Numbers round number composed	0.661	0.634
Polynomials add	0.21	0.178
Polynomials coefficient named	0.669	0.295
Polynomials collect	0.851	0.86
Polynomials compose	0.275	0.322
Polynomials evaluate	0.165	0.16
Polynomials evaluate composed	0.228	0.271
Polynomials expand	0.312	0.32
Polynomials simplify power	0.014	0.008
Probability swr p level set	0.388	0.215
Probability swr p sequence	0.293	0.167
Overall	0.54	0.52

Table 3. Accuracy on extrapolation

Module	Accuracy (Rus)	Accuracy (Eng)
Algebra polynomial roots big	0.09	0.108
Arithmetic add or sub big	0.88	0.909
Arithmetic add sub multiple longer	0.098	0.105
Arithmetic div big	0.569	0.521
Arithmetic mixed longer	0.023	0.022
Arithmetic mul big	0.332	0.323
Arithmetic mul div multiple longer	0.08	0.089
Comparison closest more	0.822	0.861
Comparison kth biggest more	0.249	0.302
Comparison sort more	0.288	0.04
Measurement conversion	0.689	0.74
Numbers place value big	0.462	0.539
Numbers round number big	0.995	0.996
Probability swr p level set more samples	0.049	0.046
Probability swr p sequence more samples	0.028	0.066
Overall	0.376	0.377

linguistic differences such as longer words, a greater number of unique characters, greater morphological variation (due to gender, case and number being grammatical categories in Russian) do not make it generally harder or easier for a Transformer to train. The modules on which the performance differs the most (polynomials coefficient named, comparison sort more) are actually the ones that were translated literally. Moreover, the performance of the model trained on the Russian version is higher on these modules.

We assume that with more training steps the performances of both models will increase by the same amount and metrics deltas will not change significantly.

5   Conclusion

We presented a translated version of DeepMind Mathematics Dataset. We adapted the linguistic templates that are used to generate the word problems to Russian. Few changes and minor additions had to be made due to the differences between English and Russian. In general, the Russian version of the dataset is equivalent to the original English one in mathematical sense and different linguistically. We also reproduced the experiment from the original paper and found out the language of the dataset does not significantly impact the performance of a Transformer model.

We release the code for dataset generation, a pre-generated dataset with 11 million word problems and the code with models training.[2]

References

1. Bender, E.M.: On achieving and evaluating language-independence in NLP. Linguist. Issues Lang. Technol. **6**, 1–26 (2011)
2. Dua, D., Wang, Y., Dasigi, P., Stanovsky, G., Singh, S., Gardner, M.: Drop: a reading comprehension benchmark requiring discrete reasoning over paragraphs. In: NAACL-HLT (2019)
3. Geva, M., Gupta, A., Berant, J.: Injecting numerical reasoning skills into language models. In: ACL (2020)
4. Hopkins, M., Bras, R.L., Petrescu-Prahova, C., Stanovsky, G., Hajishirzi, H., Koncel-Kedziorski, R.: Semeval-2019 task 10: math question answering. In: SemEval@NAACL-HLT (2019)
5. Koncel-Kedziorski, R., Hajishirzi, H., Sabharwal, A., Etzioni, O., Ang, S.D.: Parsing algebraic word problems into equations. Trans. Assoc. Comput. Linguist. **3**, 585–597 (2015)
6. Koncel-Kedziorski, R., Roy, S., Amini, A., Kushman, N., Hajishirzi, H.: Mawps: a math word problem repository. In: HLT-NAACL (2016)
7. Mielke, S.J.: Language diversity in ACL 2004–2016 (December 2016). https://sjmielke.com/acl-language-diversity.htm
8. Munro, R.: Languages at ACL 2015 (July 2015). http://www.junglelightspeed.com/languages-at-acl-this-year/

---

[2] https://github.com/mannefedov/mathematics_dataset_russian.

9. Rajpurkar, P., Zhang, J., Lopyrev, K., Liang, P.: SQuAD: 100,000+ questions for machine comprehension of text. In: EMNLP (2016)
10. Ravichander, A., Naik, A., Rosé, C.P., Hovy, E.H.: EQUATE: a benchmark evaluation framework for quantitative reasoning in natural language inference. In: CoNLL (2019)
11. Roy, S., Roth, D.: Solving general arithmetic word problems. In: EMNLP (2015)
12. Roy, S., Vieira, T., Roth, D.: Reasoning about quantities in natural language. Trans. Assoc. Comput. Linguist. **3**, 1–13 (2015)
13. Saxton, D., Grefenstette, E., Hill, F., Kohli, P.: Analysing mathematical reasoning abilities of neural models. In: International Conference on Learning Representations (2019). https://openreview.net/forum?id=H1gR5iR5FX
14. Siyam, B., Saa, A.A., Alqaryouti, O., Shaalan, K.: Arabic arithmetic word problems solver. In: ACLING (2017)
15. Trask, A., Hill, F., Reed, S.E., Rae, J.W., Dyer, C., Blunsom, P.: Neural arithmetic logic units. In: NeurIPS (2018)
16. Vaswani, A., et al.: Attention is all you need. In: NIPS (2017)
17. Wallace, E., Wang, Y., Li, S., Singh, S., Gardner, M.: Do NLP models know numbers? Probing numeracy in embeddings. In: EMNLP/IJCNLP (2019)
18. Wang, Y., Liu, X., Shi, S.: Deep neural solver for math word problems. In: Proceedings of the 2017 Conference on Empirical Methods in Natural Language Processing, pp. 845–854. Association for Computational Linguistics, Copenhagen, Denmark (September 2017). https://doi.org/10.18653/v1/D17-1088, https://www.aclweb.org/anthology/D17-1088

Searching Case Law Judgments by Using Other Judgments as a Query

Sami Sarsa[1]([✉]) and Eero Hyvönen[1,2]

[1] Semantic Computing Research Group (SeCo), Aalto University, Espoo, Finland
{sami.sarsa,eero.hyvonen}@aalto.fi
[2] HELDIG – Helsinki Centre for Digital Humanities,
University of Helsinki, Helsinki, Finland
http://seco.cs.aalto.fi, http://heldig.fi

Abstract. This paper presents an effective method for case law retrieval based on semantic document similarity and a web application for querying Finnish case law. The novelty of the work comes from the idea of using legal documents for automatic formulation of the query, including case law judgments, legal case descriptions, or other texts. The query documents may be in various formats, including image files with text content. This approach allows efficient search for similar documents without the need to specify a query string or keywords, which can be difficult in this use case. The application leverages two traditional word frequency based methods, TF-IDF and LDA, alongside two modern neural network methods, Doc2Vec and Doc2VecC. Effectiveness of the approach for document relevance ranking has been evaluated using a gold standard set of inter-document similarities. We show that a linear combination of similarities derived from the individual models provides a robust automatic similarity assessment for ranking the case law documents for retrieval.

Keywords: Legal text · Text similarity · Full text search · NLP · Document retrieval

1 Introduction: Making Case Law Search Easier

Juridical texts are widely published online by governments to make jurisdiction transparent and freely accessible to the public, organizations, and lawyers [26,28]. As juridical data, such as case law, is published, that data should also be made easily accessible. Easier access to data leads to increased transparency, as it enables more people to access the data. Additionally, by making case law search more effortless, the workload of juridical personnel can be reduced, leading to savings in litigation costs.

This paper makes a step towards easy public access for juridical data by presenting an effective case law search method using other judgments or free text as query input. A publicly available web application for querying Finnish case law is presented to evaluate the method. The efficiency, i.e., achieving maximum

A. Filchenkov et al. (Eds.): AINL 2020, CCIS 1292, pp. 145–157, 2020.
https://doi.org/10.1007/978-3-030-59082-6_11

productivity with minimum effort, is improved compared to traditional keyword based querying by allowing uploading of case law files. In addition, the application provides "Get similar" buttons for retrieval results enabled by a simple API that returns similar cases given an ECLI identifier [13]. This removes the need to come up with relevant keywords and allows fast exploratory search with meaningful results as queries. The API also enables use of the application's similarity computation programmatically for research purposes or for use in other applications. The prototype application has been included as a use-case application perspective in the LawSampo semantic portal [16].

Besides efficiency, document retrieval effectiveness is required to be at least satisfactory in order to improve overall document retrieval. The main concern in retrieval effectiveness is the ranking of retrievable documents based on relevance to a query [10,25]. Using the assumption underlying vector space models, i.e., that "the relevance of a set retrieved documents to a query is approximately equal to similarity between the query and documents in retrieved set." [15], a desirable ranking can be obtained by sorting computed correlations of texts' vector representations. The vector representations are referred to as "embeddings", since the texts are embedded into a vector space. Our application combines traditional word frequency based text embedding models with newer neural network based models to provide meaningful textual similarity rankings that are able to take synonyms and other word relations into account.

In the following, the method and its prototype implementation are first described. After this, evaluation results of the underlying methods are presented. In conclusion, contributions of our experiments and related work are discussed.

2 The Finnish Case Law Finder Application

Data. The Finnish case law corpus for the application is provided by the Finnish Ministry of Justice as part of Semantic Finlex data service[1] [27]. This Finnish case law corpus consists of 13 053 judgments from 1980 to 2019 at the moment. The case law texts contain references to the laws that are applied in giving the legal decisions. This helps automatic similarity computation, since judgments that have one or more applied laws in common inherently convey that they are meaningfully similar to each other. However, the laws appear in text in either abbreviated or in their full form making their identification difficult. To harmonize the texts we use regular expressions to expand the abbreviations as a preprocessing step. Another hindrance for embedding models is word inflections. Finnish language is agglutinative, which causes words to often appear in multiple forms. This leads to a large vocabulary and small frequencies for different words making it difficult to automatically infer relationships between the words with limited data. We reduce the effect of word inflections on embedding models by using LAS [21] to lemmatize, i.e., to normalize inflected words to their base form, before embedding texts for similarity computation. In addition, we filter out

---

[1] https://data.finlex.fi.

stopwords from the case law texts as this has been shown to improve document retrieval [7,34].

The case law data is stored in a relational database. It contains a table for documents that includes document texts, metadata, and an integer document identifier. The identifier corresponds to the document's index in training data for embedding models and is used to retrieve documents. The database also includes tables for users and similarity ratings to enable users to rate document pair similarities within the application. User-rated similarities are used to evaluate the application's effectiveness.

Similarity Computation. Our application ranks documents for retrieval by sorting similarity values that are obtained by computing the correlation of the texts' vector representations. This is similar to the vector space model [33] that remains widely used [5]. We chose the standard method [22], cosine similarity as the application's correlation measure for text embedding similarity.

For embedding generation, we selected four models. Two of these are bag-of-words models, i.e., word frequency based models, namely TF-IDF [36] and LDA [6]. The other two models, Doc2Vec [20] and Doc2VecC [12], represent more modern text embedding methods: they are extensions to the word embedding neural network model Word2Vec [24] that is able to map words' semantic meanings close to each other when trained with large amounts of texts. Like Word2Vec, Doc2Vec and Doc2VecC are neural networks that learn vector representations by learning to either predict missing words from context, or to predict context words given a single word.

As our models are different in nature, we created a weighted ensemble of the models to improve upon the individual models' effectiveness in producing text embeddings for ranking. Multi-co-training TF-IDF, LDA, and Doc2Vec has been shown to outperform the individual models in topic classification [18]. However, unlike topic classification, our task enables us to use a simpler approach to create ensembles of the models. Our goal is to infer real-valued similarities from texts instead of classifying the texts. Hence, we construct our ensembles models with minimal effort by computing weighted averages over the cosine similarities from the individual models' embeddings.

We obtain weights for similarities from different models' embeddings using linear regression presented in Eq. (1)

$$y = x_1\beta_1 + .. + x_n\beta_n + \epsilon = \boldsymbol{x}^T\boldsymbol{\beta} + \epsilon, \tag{1}$$

where $\epsilon \in \mathbb{R}$ is an error term denoting disturbance in the linear relation. Linear regression assumes its inputs $\boldsymbol{x} \in \mathbb{R}^n$ are linearly related to an observed variable $y \in \mathbb{R}$. In our case, $\boldsymbol{x}$ contains similarity values for a document pair computed from embeddings given by the individual models and y is a ground truth human assigned similarity value for the pair of case law documents.

Full Document as a Query. The main goal of the application is to enable efficient and precise search with full texts. Full texts, however, come in various formats. For instance, the user might have a case law text in print or a file in some text format such as XML or plain text. Reading text content from a text file is straightforward, but analyzing a photographed image of printed text, or a PDF with text as image, requires a technique called optical character recognition (OCR). Thus, we included the Tesseract OCR [35] application in our web application to enable querying case law with photographed texts. Tesseract OCR was chosen for the task because it is an open source OCR system that has a well performing pre-trained model for Finnish text, comprehensive documentation, and the possibility of retraining the model further. Although having a model for Finnish out-of-the-box, Tesseract OCR was not directly implemented into the software. Instead, it was first retrained to include letters "Å", "å" and the section sign "§", which were not included in the Tesseract OCR's readily available Finnish text model.

Figure 1 depicts the end-user interface of the application, *Semantic Finlex case law finder*. The user is able to input a text document as a query to Finlex case law either by uploading a file or by writing text directly to the form. The query document is seen on the sub-window "Document content". Supported file formats for uploading documents are plain text, XML, PDF, and with Tesseract OCR, image formats, such as JPEG or PNG. The text extraction mode to be used can be selected by the drop-down menu on the right bottom corner. The search form also allows the user to choose the algorithm that ranks the documents by using the drop-down menu on the left bottom in Fig. 1. Here the method "Ensemble" is selected. Ranking with some algorithms may work better than others for certain topics, or depending on what kind of relatedness is preferred. Also the preferred result size can be specified.

Fig. 1. Semantic Finlex case law finder application document search.

Document Ranking. Once a document is submitted in the document search form, it is sent to the application back-end that handles document ranking. The back-end provides a simple API for retrieving case law documents. The query document is sent to the API via a HTTP/POST request where an embedding model is specified in the requested URI. An optional parameter n is provided to limit the number of retrieved documents, since sending the results via HTTP causes a bottleneck in retrieval time. The query document is pre-processed to the same format as models' training data, and the formatted query text is given to the model as input. The model transforms its input into a vector, and cosine similarity values are computed between the query's vector representation and all case law documents' vector representations in the underlying database. Then, all document ids are sorted by the computed similarities, and the top n ranked documents are retrieved from the database and returned in JSON-format. The query retrieval and processing is illustrated as a graph in Fig. 2.

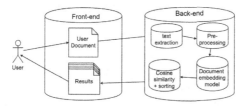

Fig. 2. LawSampo case law finder application architecture overview.

The ranked cases are shown to the user as a list of expandable panels that have the case law identifier and keywords as a panel's title, as depicted in Fig. 3. Similarity rating is shown by default to give insight on how the similarities change. This allows the user to see from the values whether there are likely relevant results left to view. The result items also contain the button "GET SIMILAR" for quickly querying case law related to a result document. This is intended as a helpful measure when the user does not have a certain query document, but rather wants to search the case law corpus exploratively. The retrieved results may also be further filtered by traditional methods, e.g. by using exact phrase match or filtering by court.

The created application and its ranking models are intended to work as generally as possible. While the application is created for the specific domain of Finnish case law, there is little in the implementation, besides the trained models, that restricts the application to be used with other corpora. Abbreviation expansion is the only element that ties the textual context to the Finnish language or juridical terminology. This suggests that the performance can be generalized to texts from other linguistic domains as well as other languages.

The models in the application leverage lemmatization as it was deemed beneficial in model evaluation. Additionally, by effectively normalizing word inflections, lemmatization also helps automatic inference of natural sentence queries.

Fig. 3. Document search results in Semantic Finlex case law finder.

Thus, we performed an additional free text query test on our working application with short natural language queries to see how it would manage the task although the models are optimized for full document search. A working example is the query "törmäsin autoon" (I collided into a car). Without lemmatization, the inflected word "törmäsin" (I collided) would be non-existent in the training data. This would prevent even the neural net models from inferring that the user is inquiring about collisions. As a result of the free text query experiment, we find that the example query accurately returns cases concerning cars and traffic accidents. However, we did not perform any extensive free text query testing as this was not our primary objective.

3   Evaluation

Method. We assessed ranking effectiveness of the four text embedding models that are used in the application, namely of TF-IDF, LDA, Doc2Vec and Doc2VecC. Additionally, we tested using word vector averages of Word2Vec as text embeddings. The embedding effectiveness is evaluated by comparing inter-document similarities computed from the embeddings against gold standard similarities. For our performance measures, we use Pearson correlation, Spearman rank order correlation and mean squared error. We tested different preprocessing methods effect on the models. Most of the models contain manually assignable hyperparameters that we tuned using the gold standard. For preprocessing steps, we tested the effects of lemmatization and stopword removal. Additionally, query expansion was tested with the word-frequency models TF-IDF and LDA using

OIKO,[2] an ontology of Finnish legal terms, and a Finnish ontology collection KOKO[3] to find synonyms, hypernyms and hyponyms for document words.

Gold Standard Labels.

As ground truth data for ranking effectiveness evaluation, we use manually labeled, i.e. gold standard, similarities for selected document pairs. However, gold standard similarity scores are strenuous to obtain and can be the most resource consuming step in the creation of an information retrieval system [30]. In addition, determining how to judge document similarity might not be obvious for an annotator regardless of their expertise. Probably due to this, people tend to have different opinions on similarity [1,2]. Therefore, it is reasonable to devise an intelligent system that is easy to use, and which leverages preferably more than one annotator to acquire similarity labels for a gold standard set.

To alleviate the manual similarity labelling process, we incorporate comparing and evaluating case law similarity within our self-built case law finder web application. Labeling inter-document similarity within the application is made possible by providing an optional sign up and login, which allows user to submit a similarity assessment for a query result. We include sign up and login requirement to select verified assessors from others, as well as to exclude thoroughly inconsistent similarity labels, i.e., random or seemingly dishonest submissions.

Fig. 4. Rating similarity in the Finnish case law finder application. Green tick indicates a query-result pair similarity has been rated. (Color figure online)

Similarly to SemEval's methodology for acquiring ground truth similarities [1,11,23], we use a 0–5 scale for similarity: Almost identical (5); Similar topics and content (4); Multiple shared topics (3); At least one common topic (2); Some

---

[2] https://finto.fi/oiko/en/.
[3] https://finto.fi/koko/en/.

common elements (1); Completely different (0). Documents for the gold standard set of inter-document similarities were selected by choosing seven "seed" documents of different topics including crime, accidental homicide, nature preservation, and name change. Then the application was used to search for documents similar to the seeds and the first results were rated. The final similarity rating set obtained includes 138 distinct ratings in total. Due to difficulty in acquiring expert labels (likely a result of lack of advertising or not providing a materialistic reward for rating), 129 of the labels are assigned by the first author of this paper, while nine are given by a volunteering law student.

As the gold standard is constructed mostly by a single person and not someone more familiar with case law, or better yet, multiple such people, our ground truth for evaluating Finnish case law ranking leaves plenty of room for improvement. However, in our defence, 2017 SemEval task [11] in multi-lingual text similarity uses 250 pairs for each language, which are either constructed by a single expert or non-experts via crowd-sourcing. Additionally, regarding the gold standard set size, Campr et al. [9] had three annotators to manually label only 150 pairs of summaries in order to compare the accuracy of various similarity computation models.

Results. Consistently with all our selected assessment measures, Doc2Vec and LDA perform the best of the individual models with correlation measures between 0.60 and 0.65. TF-IDF and Doc2VecC also produce decent correlations with values below 0.6 but clearly above 0.5 for both. For Word2Vec averages, the results are not promising as its performance was far below the other models. While there are differences between the individual model performances, they are not clear enough to provide conclusive evidence for supremacy of the best performing individual models, especially when taking into account our inextensive validation set. However, both mean ensemble and linear regression ensemble outperform the individual models significantly. This shows that leveraging multiple models is recommendable when computing similarities for ranking using embedding models although it requires a bit of extra work to train all the models. The results are shown in Table 1. Regarding preprocessing steps, lemmatization is found useful for all models. For stopword removal, we found initially mixing results for its usefulness. However, after keeping the stopwords "yli" (over) and "ei" (no) in the texts while removing other stopwords, we found that stopword removal was beneficial for all models. LDA benefited from query expansion for training data while TF-IDF was unaffected.

Table 1. Evaluation results. Correlations and mean squared error between gold standard similarities and embeddings' cosine similarity values for best hyperparameters and preprocessing steps. Results for embeddings from lemmatized texts are without brackets and ones without lemmatization in round brackets. Optimal embedding size, model variant window size and random sampling rate hyperparameters are shown for machine learning models where applicable. The slash symbol "/" between two options denotes the options resulting in equal performance for model evaluation.

Model	Pearson	Spearman	MSE	Embedding size	Variant	Window
TF-IDF	0.57 (0.42)	0.56 (0.50)	0.85 (1.17)	-	-	-
LDA	0.62 (0.46)	0.60 (0.48)	0.76 (1.07)	300	-	-
Word2Vec	0.42 (0.54)	0.41 (0.52)	1.16 (0.93)	900	C-BoW	10
Doc2Vec	0.64 (0.48)	0.64 (0.46)	0.71 (1.03)	500	D-BoW	5/10
Doc2VecC	0.56 (0.53)	0.53 (0.52)	0.89 (0.94)	700	C-BoW	10
Mean ensemble	0.70 (0.62)	0.69 (0.62)	0.61 (0.76)	-	-	-
LinReg ensemble	0.75 (0.70)	0.74 (0.70)	0.5 (0.6)	-	-	-

Since our best performing model is linear regression ensemble, which computes weights for each individual model, we tested the validity of the weights using fivefold cross-validation. The results, depicted in Table 2, show that for the test set, linear regression ensemble performs similarly to the mean ensemble. This suggests that more ground truth data is required to optimize the weights, but also that fine-tuning the weights might not carry much importance. We further analysed the weights to see which models are deemed the most important by the regression. By examining the regression weights in Table 3, we see that Doc2Vec contributes to approximately a third of the ensemble's similarity score. This implies that neural networks are a bigger factor in the ensembles success than the more traditional bag-of-words models. On the other hand, according to the weights, Word2Vec contributes nothing in addition to its two derivatives, and thus, is not needed in the ensemble.

Table 2. Average test set correlations and mean squared error between gold standard similarities and embeddings' cosine similarity values for fivefold cross validation.

Model	Pearson	Spearman	MSE
TF-IDF	0.47	0.37	1.06
LDA	0.58	0.51	0.83
Word2Vec	0.44	0.40	1.13
Doc2Vec	0.58	0.58	0.84
Doc2VecC	0.59	0.51	0.81
Mean ensemble	0.68	0.65	0.64
LinReg ensemble	0.68	0.65	0.64

Table 3. Average linear regression weights from fivefold cross-validation models

TF-IDF	LDA	Doc2Vec	Doc2VecC	Word2Vec
0.22	0.10	0.41	0.27	0.00

4   Contributions and Related Work

We present a public web application for efficient and effective retrieval of Finnish case law documents. The application improves upon traditional document retrieval efficiency by introducing the possibility to use documents of various formats as the query text. The general idea of searching documents similar to a description in natural language text is not new: it has been applied in situations where rich textual target documents are available, formulating the query is challenging in terms of keywords, and there is a similar document available to be used as the model to be matched. This is often the case with legal data, and there are commercial systems on the Web, such as Casetext[4] Fastcase,[5] for searching legal documents using other documents as a query. The idea has been applied also in, for example, patent search[6] where a patent application or a description of it can be used for checking out whether the idea has already been patented. However, we have not been able to find research publications describing the benefits of providing easier case law search by making querying easier using documents, or about public freely available applications with such goals. Our work is an attempt to fill this gap, and to create an open source implementation and data for the task (CC-BY-4.0).

Our presented application's effectiveness is based on combining existing text embedding models for similarity computation. We evaluated TF-IDF, LDA, Doc2Vec, Doc2VecC and Word2Vec averages as embedding models for retrieval ranking against gold standard similarities. Our results show that a linear regression ensemble and also a simple mean weighted ensemble are more powerful than individual models. Based on our results, we propose using different types of models to compute inter-document similarity although this requires this the extra work to train all of them. Even two highly similar neural network models Doc2Vec and Doc2vecC both appear to contribute to ensemble performance. However, using Word2Vec averages with the two more advanced models is redundant in the light of our results. While our results show that mean weights provide significant increase in model performance over individual models, provided that one has a gold standard set, we suggest using linear regression or other learning method to learn optimal weighting for the models. We must note that our gold standard set is small and thus different results, especially for individual models and also linear regression weights, are possible given more training or evaluation data.

---

[4] http://casetext.com.
[5] http://fastcase.com.
[6] Cf., e.g., http://www.acclaimip.com/.

Text embeddings by neural networks has gained attention in recent work on information retrieval [4] and textual semantic similarity [11,23]. However, the focus in semantic similarity computation has been on short texts, such as keyword queries or sentences. In contrast, the Finnish case law data used in this work consists of variously sized documents, the longer ones containing over 10 000 words. As for the legal context, neural network embeddings have been leveraged in mildly related cases. For instance, Ash et al. [3] analyse judges' relations and judicial reasoning by examining spacial relationships between case law embeddings of different judges' verdicts. Moreover, closely related to case law retrieval, Landthaler et al. [19] have used word embeddings to enhance retrieval of EU Data Protection Directives.

Future work would include studying the applicability of more novel neural network models such as ELMO [29], BERT [14], and ones from the GPT-family [8,31,32]. The new models are mainly developed for using massive amounts of training data, and thus, the possible benefits of transfer learning with non-legal texts should be investigated. Since the performance evaluation performed in this work is far from optimal, future work would also include testing models with more robust evaluation with a better gold standard set and more information retrieval metrics, e.g. cumulative gain measures [17].

Acknowledgements. Thanks for collaborations to Aki Hietanen, Saara Packalen, Tiina Husso, and Oili Salminen at Ministry of Justice, Finland, to Minna Tamper and Jouni Tuominen at Aalto University and University of Helsinki, and to Jari Linhala, Arttu Oksanen, and Risto Talo and at Edita Publishing Ltd. This work is part of the Anoppi project funded by Finnish Ministry of Justice (https://oikeusministerio.fi/en/project?tunnus=OM042:00/2018).

References

1. Agirre, E., Cer, D., Diab, M., Gonzalez-Agirre, A., Guo, W.: * SEM 2013 shared task: semantic textual similarity. In: Second Joint Conference on Lexical and Computational Semantics (* SEM), Volume 1: Proceedings of the Main Conference and the Shared Task: Semantic Textual Similarity, vol. 1, pp. 32–43 (2013)
2. Agirre, E., Diab, M., Cer, D., Gonzalez-Agirre, A.: Semeval-2012 task 6: a pilot on semantic textual similarity. In: Proceedings of the First Joint Conference on Lexical and Computational Semantics-Volume 1: Proceedings of the main conference and the shared task, and Volume 2: Proceedings of the Sixth International Workshop on Semantic Evaluation, pp. 385–393. Association for Computational Linguistics (2012)
3. Ash, E., Chen, D.L.: Case vectors: spatial representations of the law using document embeddings. Social Science Research Network (Working paper) (2018)
4. Basu, M., Ghosh, S., Ghosh, K.: Overview of the fire 2018 track: information retrieval from microblogs during disasters (IRMiDis). In: Proceedings of the 10th Annual Meeting of the Forum for Information Retrieval Evaluation, FIRE 2018, pp. 1–5. ACM, New York (2018)
5. Beel, J., Gipp, B., Langer, S., Breitinger, C.: Research-paper recommender systems: a literature survey. Int. J. Digit. Libr. **17**(4), 305–338 (2015). https://doi.org/10.1007/s00799-015-0156-0

6. Blei, D.M., Ng, A.Y., Jordan, M.I.: Latent Dirichlet allocation. J. Mach. Learn. Res. **3**, 993–1022 (2003)
7. Brants, T.: Natural language processing in information retrieval. In: Proceedings of the 14th Meeting of Computational Linguistics in the Netherlands, pp. 1–12 (2004)
8. Brown, T.B.: Language models are few-shot learners. arXiv preprint arXiv:2005.14165 (2020)
9. Campr, M., Ježek, K.: Comparing semantic models for evaluating automatic document summarization. In: Král, P., Matoušek, V. (eds.) TSD 2015. LNCS (LNAI), vol. 9302, pp. 252–260. Springer, Cham (2015). https://doi.org/10.1007/978-3-319-24033-6_29
10. Cao, Y., Xu, J., Liu, T.-Y., Li, H., Huang, Y., Hon, H.-W.: Adapting ranking SVM to document retrieval. In: Proceedings of the 29th Annual International ACM SIGIR Conference on Research and Development in Information Retrieval, pp. 186–193. ACM (2006)
11. Cer, D., Diab, M., Agirre, E., Lopez-Gazpio, I., Specia, L.: Semeval-2017 task 1: semantic textual similarity-multilingual and cross-lingual focused evaluation. arXiv preprint arXiv:1708.00055 (2017)
12. Chen, M.: Efficient vector representation for documents through corruption. In: 5th International Conference on Learning Representations. OpenReview.net (2017)
13. Council of the European Union: Council conclusions inviting the introduction of the European Case Law Identifier (ECLI) and a minimum set of uniform metadata for case law. In: Official Journal of the European Union, C 127, 29.4.2011, pp. 1–7. Publications Office of the European Union (2011)
14. Devlin, J., Chang, M.-W., Lee, K., Toutanova, K.: Bert: Pre-training of deep bidirectional transformers for language understanding. arXiv preprint arXiv:1810.04805 (2018)
15. Pandey, S., Purohit, G.N., Munshi, U.M.: Data security in cloud-based applications. In: Munshi, U.M., Verma, N. (eds.) Data Science Landscape. SBD, vol. 38, pp. 321–326. Springer, Singapore (2018). https://doi.org/10.1007/978-981-10-7515-5_24
16. Hyvönen, E., et al.: LawSampo: a semantic portal on a linked open data service for Finnish legislation and case law. In: Proceedings of ESWC 2020, Poster and Demo Papers. Springer, Heidelberg (2020, in press)
17. Järvelin, K., Kekäläinen, J.: Cumulated gain-based evaluation of IR techniques. ACM Trans. Inf. Syst. (TOIS) **20**(4), 422–446 (2002)
18. Kim, D., Seo, D., Cho, S., Kang, P.: Multi-co-training for document classification using various document representations: TF-IDF, LDA, and Doc2Vec. Inf. Sci. **477**, 15–29 (2019)
19. Landthaler, J., Waltl, B., Holl, P., Matthes, F.: Extending full text search for legal document collections using word embeddings. In: JURIX, pp. 73–82 (2016)
20. Le, Q., Mikolov, T.: Distributed representations of sentences and documents. In: International Conference on Machine Learning, pp. 1188–1196 (2014)
21. Mäkelä, E.: LAS: an integrated language analysis tool for multiple languages. J. Open Source Softw. **1**(6), 35 (2016)
22. Manning, C.D., Raghavan, P., Schütze, H.: Introduction to Information Retrieval, Chap. 6. Cambridge University Press, New York, NY, USA (2008)

23. Marelli, M., Bentivogli, L., Baroni, M., Bernardi, R., Menini, S., Zamparelli, R.: Semeval-2014 task 1: evaluation of compositional distributional semantic models on full sentences through semantic relatedness and textual entailment. In: Proceedings of the 8th International Workshop on Semantic Evaluation, SemEval 2014, pp. 1–8 (2014)

24. Mikolov, T., Sutskever, I., Chen, K., Corrado, G., Dean, J.: Distributed representations of words and phrases and their compositionality. CoRR, abs/1310.4546 (2013)

25. Nalisnick, E., Mitra, B., Craswell, N., Caruana, R.: Improving document ranking with dual word embeddings. In: Proceedings of the 25th International Conference Companion on World Wide Web, WWW 2016 Companion, pp. 83–84. International World Wide Web Conferences Steering Committee, Republic and Canton of Geneva, Switzerland (2016)

26. Oksanen, A., Tuominen, J., Mäkelä, E., Tamper, M., Hietanen, A., Hyvönen, E.: Semantic Finlex: Finnish legislation and case law as a linked open data service. In: Proceedings of Law via the Internet 2018: Knowledge of the Law in the Big Data Age (abstracts), LVI 2018, pp. 212–228 (October 2018)

27. Oksanen, A., Tuominen, J., Mäkelä, E., Tamper, M., Hietanen, A., Hyvönen, E.: Semantic Finlex: transforming, publishing, and using Finnish legislation and case law as linked open data on the web. In: Peruginelli, G., Faro, S. (eds.) Knowledge of the Law in the Big Data Age. Frontiers in Artificial Intelligence and Applications, vol. 317, pp. 212–228. IOS Press (2019). ISBN 978-1-61499-984-3 (print); ISBN 978-1-61499-985-0 (online)

28. van Opijnen, M., Peruginelli, G., Kefali, E., Palmirani, M.: On-line publication of court decisions in the EU: report of the policy group of the project 'building on the European case law identifier' (15 February 2017). https://ssrn.com/abstract=3088495, http://dx.doi.org/10.2139/ssrn.3088495

29. Peters, M.E., et al.: Deep contextualized word representations. In: Proceedings of NAACL (2018)

30. Qin, T., Liu, T.-Y., Xu, J., Li, H.: LETOR: a benchmark collection for research on learning to rank for information retrieval. Inf. Retr. **13**(4), 346–374 (2010)

31. Radford, A., Narasimhan, K., Salimans, T., Sutskever, I.: Improving language understanding by generative pre-training (2018)

32. Radford, A., Wu, J., Child, R., Luan, D., Amodei, D., Sutskever, I.: Language models are unsupervised multitask learners. OpenAI Blog **1**(8), 9 (2019)

33. Salton, G., Wong, A., Yang, C.S.: A vector space model for automatic indexing. Commun. ACM **18**(11), 613–620 (1975)

34. Shin, J.-H., Abebe, M., Yoo, C.J., Kim, S., Lee, J.H., Yoo, H.-K.: Evaluating the effectiveness of the vector space retrieval model indexing. In: Park, J.J.J.H., Pan, Y., Yi, G., Loia, V. (eds.) CSA/CUTE/UCAWSN-2016. LNEE, vol. 421, pp. 680–685. Springer, Singapore (2017). https://doi.org/10.1007/978-981-10-3023-9_104

35. Smith, R.: An overview of the Tesseract OCR engine. In: Proceedings of the Ninth International Conference on Document Analysis and Recognition, ICDAR 2007, vol. 2, pp. 629–633. IEEE Computer Society, Washington, DC, USA (2007)

36. Sparck Jones, K.: A statistical interpretation of term specificity and its application in retrieval. J. Doc. **28**(1), 11–21 (1972)

GenPR: Generative PageRank Framework for Semi-supervised Learning on Citation Graphs

Mikhail Kamalov$^{(\boxtimes)}$ and Konstantin Avrachenkov

Inria Sophia Antipolis, Valbonne, France
{mikhail.kamalov,k.avrachenkov}@inria.fr

Abstract. Nowadays, Semi-Supervised Learning (SSL) on citation graph data sets is a rapidly growing area of research. However, the recently proposed graph-based SSL algorithms use a default adjacency matrix with binary weights on edges (citations), that causes a loss of the nodes (papers) similarity information. In this work, therefore, we propose a framework focused on embedding PageRank SSL in a generative model. This framework allows one to do joint training of nodes latent space representation and label spreading through the reweighted adjacency matrix by node similarities in the latent space. We explain that a generative model can improve accuracy and reduce the number of iteration steps for PageRank SSL. Moreover, we show that our framework outperforms the best graph-based SSL algorithms on four public citation graph data sets and improves the interpretability of classification results.

Keywords: Semi-supervised learning · Generative model · PageRank · Citation graphs · Neural networks

1 Introduction

The main idea of SSL is to solve a classification task with an extremely low number n_l of labeled data points in comparison with the number n_u of unlabeled data points ($n_l \ll n_u$). Therefore, with regard to citation graphs with a huge amount of nodes (e.g. Pubmed, MS Academic) SSL is a good technique to avoid preparing data points for supervised learning. The area of SSL focusing on the classification of nodes in citation graphs, in particular, citation graphs is called a graph-based SSL. The standard input for graph-based SSL algorithms is a graph $\mathcal{G} = (\mathcal{V}, \mathcal{E})$ with $n = n_l + n_u = |\mathcal{V}|$ nodes (papers), $e = |\mathcal{E}|$ edges (citations), $A \in \mathbb{R}^{n \times n}$ is the adjacency matrix and $X = (x_{i,j})_{i,j=1}^{n,d}$ is a matrix of nodes where each node $x_i = (x_{i,1}, \ldots, x_{i,d}) \in \mathbb{R}^d$ has a feature representation in d-space. In the context of citation graphs X is a bag-of-words representation for the nodes (papers). Moreover, each node belongs to one of c classes $\{\mathcal{C}_1, \ldots, \mathcal{C}_c\}$. Also we

This work was run in the frame of a 3IA labeled MyDataModels INRIA joint project, supported by MyDataModels via an I-Lab research grant.

A. Filchenkov et al. (Eds.): AINL 2020, CCIS 1292, pp. 158–165, 2020.
https://doi.org/10.1007/978-3-030-59082-6_12

have the labels matrix $Y = (y_{i,j})_{i,j=1}^{n,c} \in \mathbb{R}^{n \times c}$ such that $y_{i,j} = 1$ if $x_i \in C_j$ and $y_{i,j} = 0$ otherwise. Nowadays, the area of the graph-based SSL consists of two main research directions:

- the classical diffusion-based linear algorithms that spread the class information through the adjacency matrix A: Label Propagation (LP) [9], PageRank SSL (PRSSL) [1];
- the graph convolution-based neural network (NN) algorithms that apply the dot product of adjacency matrix A with NN nonlinear transformation of features X for the classification. The recently proposed: approximated Personalized graph NN (APPNP) [6], Graph Attention Network (GAT)[8], Graph Convolution Network (GCN) [5].

Regarding graph-based SSL algorithms, one can notice that they use A with binary weights on edges (citations), which can cause a loss of information about node similarities and further may negatively affect label diffusion through A. We address this issue through the following contributions:

- We propose a novel graph-based SSL inductive (I)/ transductive (T) framework, created by embedding PRSSL [1] in generative model (GenPR);
- We show that the generative model can be used to reweight A to further improve PRSSL label spreading;
- We show that GenPR improves the interpretability of NN classification results based on the information about nodes similarity in the latent space;
- We show that GenPR outperforms the recently proposed algorithms for graph-based SSL and reduces the number of steps of PageRank [7] to obtain more accurate classification results.

2   Related Work

Our framework is based on the combination of the following two ideas:

1. PRSSL [1] gives a PowerIteration based explicit solution for the graph classification: $F^t = \alpha D^{-\sigma} A D^{\sigma-1} F^{t-1} + (1-\alpha)Y$; $t \geq 0$ where F^t is a result of the t-th iteration and α is a regularization parameter in the range $[0, 1]$ and σ is a power of $D_{i,i} = \sum_{j=1}^{n} A_{i,j}$;
2. generative semi-supervised model (M2) [4] : $p(x, \hat{y}, z) = p(x|z, \hat{y})p(\hat{y})p(z)$ where $p(z) = \mathcal{N}(z|0, I)$, $p(\hat{y})$ is a categorical distribution of latent class variable and $p(x|z, \hat{y})$ is a nonlinear transformation of the latent variables z and $\hat{y}$.

Since our framework is the graph convolution-based NN algorithm with PageRank, we need to define the main difference with APPNP [6]. The difference is that GenPR jointly trains a redefined generative model [4] and PRSSL [1] with a linear combination of A and similarity matrix in latent space, while APPNP applies PageRank with default A as a preprocessing step for output of multilayer perceptron (MLP).

3   Generative PageRank (GenPR)

3.1   Intuition of GenPR

Before we go into details of our framework let us define the motivation and the intuition behind GenPR. The main idea of GenPR is to resolve the following issues:

1. $A_{i,j} = 1$ does not provide the information about impact of cited paper j on the citing one i;
2. $A_{i,j} = 0$ may show that author i did not cite the paper j, but he could have used some information from it.

Let us define some useful notation for the GenPR intuition: $x_i \in X$ is a i.i.d. samples of some continuous random variable x, then an output of MLP $Y^* = (y_i^*)_{i=1}^n \in \mathbb{R}^{n \times c}$ is a sample from random variable y^* given x as an input; $Z = (z_i)_{i=1}^n \in \mathbb{R}^{n \times d'}$ where z_i is a latent representations of each node x_i sampled from latent random variable z in d'-space; $W = (w_{i,j})_{i,j=1}^n \in \mathbb{R}^{n \times n}$ is a similarity matrix where each element $w_{i,j} = h(z_i, z_j)$, $\forall z_i, z_j \in Z$ is an output of some positively defined kernel h; $A' = A + \gamma W$ is a reweighted adjacency matrix A with a parameter $\gamma \in [0, 1]$ for W; $D'_{i,i} = \sum_{j=1}^n A'_{i,j}$ is a diagonal matrix.

Let us redefine the recurrent formula of PRSSL [1] using Y^* at each training epoch as a replacement of real labels $Y = Y^*$:

$$F^t = \alpha D'^{(-\sigma)} A' D'^{(\sigma-1)} F^{t-1} + (1-\alpha) Y^*; \tag{1}$$

where $F^t = (y_{i,j}^{pr})_{i,j=1}^{n,c} \in \mathbb{R}^{n \times c}$. Here $y_i^{pr} = (y_{i,1}^{pr}, \ldots, y_{i,c}^{pr})$ is a sample from random variable y^{pr} since (1) is a transformation of the random variable y^* and $F^0 = Y^*$.

Then assume that F^t will improve the accuracy of Y^* by using the information of nodes similarity in latent space during the t-th iterations. We named it the PageRank spreading assumption. Moreover, we propose to use Y^* as a new labels. Let us notice that and $y^* \sim y^{pr}$ due to the PowerIteration PageRank property $||F^t - Y^*||_1 \leq \frac{1}{1-\alpha}||F^1 - Y^*||_1$ [2](Property 12). This allows us consistently use the aforementioned PageRank spreading assumption in training process of the generative model:

$$p(x, y^*, z) \approx p(x|z, y^{pr})p(y^{pr})p(z) \tag{2}$$

where $p(\cdot)$ is a PDF of a random variable.

3.2   Objective Function of GenPR

In this subsection we consider the inductive regime of GenPR which allows us to train jointly the generative model (2) and PRSSL (1). Now let us define GenPR objective function. It is obtained by maximizing the variational lower bound of the data log-likelihood of (2) with variance ϕ and generative θ parameters [3]:

$$\log p(x, y^*) \geq \mathbb{E}_{q_\phi(z|x,y^*)}\big[\log p_\theta(x|z, y^{pr})\big]$$
$$+ \mathbb{E}_{q_\phi(z|x,y^*)}\big[\log p_\theta(y^{pr})\big] - D_{KL}(p(z)||q_\phi(z|x, y^*)) \tag{3}$$

where $q_\phi(z|y^*, x) = \mathcal{N}(z|\mu(y^*, x), \sigma^2(x))$ is a multivariate Gaussian distribution parameterized by $\mu(y^*, x)$ and $\sigma(x)$ that are inferred from NN layers for expectation and variance respectively; $p_\theta(x|z, y^{pr}) = f_\theta(z, y^{pr})$ is a nonlinear transformation of z and y^{pr} by NN layer; $p_\theta(y^{pr}) = PR(y^*, \mu(y^*, x), A)$ is a linear transformation of y^* by (1) (the NN layer version will be defined in the next subsection 3.3), $p(z) = \mathcal{N}(z|0, I)$ is a multivariate Gaussian distribution and $D_{KL}(\cdot||\cdot)$ is the Kullback-Leibler divergence.

Since we can trade the quality generation of x for the quality of y_i^{pr} and estimate y_i^{pr} using the information from n_l, we can use $\beta \in [0,1]$ as a weight parameter for $p_\theta(x|z, y^{pr})$ and the categorical crossentropy $\mathcal{U}(F^t, Y) = \sum_{i=1}^{n_l} \sum_{j=1}^{c} (y_{i,j} \cdot \log(y_{i,j}^{pr}))$ for y_i^{pr} estimation. Thus, we obtain from (3) the final inductive (I) GenPR objective function:

$$\mathcal{L}(\theta, \phi, x, Y) = \beta \mathbb{E}_{q_\phi(z|x,y^*)} \big[\log p_\theta(x|z, y^{pr}) \big] + \log p_\theta(y^{pr})$$
$$- D_{KL}(p(z)||q_\phi(z|x, y^*)) - \mathcal{U}(F^t, Y) \tag{4}$$

The difference between inductive (I) and transductive (T) regimes of GenPR is that transductive GenPR does not use the proposition that y^* is a new labels and an objective function looks as follows:

$$\mathcal{L}_T(\theta, \phi, x, Y) = \beta \mathbb{E}_{q_\phi(z|x)} \big[\log p_\theta(x|z) \big]$$
$$- D_{KL}(p(z)||q_\phi(z|x)) - \mathcal{U}(F^t, Y) \tag{5}$$

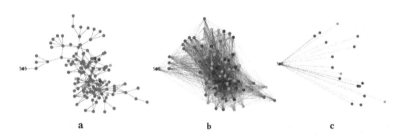

Fig. 1. Sample nodes from Citeseer data set: a - A before GenPR, where colored nodes are labeled and grey are unlabeled, straight black edges are citations between nodes (papers); b - A' after GenPR, where all colored nodes are result from F^t, and color of an edges by weights from A' (cyan is a lower weights, maroon is a higher weights); c - the result of filtering lower weight edges for the node 545.

3.3   Architecture of GenPR

Since we have defined the objective function of GenPR (4) we can explain the GenPR layers architecture. The part of z inference contains the following layers:

$$Y^* = \pi_\theta(X); \quad \pi_\theta(X) = h_1(XW_1 + B_1) \tag{6}$$

$$\mu(X, Y^*) = h_\mu(concat(X, Y^*)W_\mu + B_\mu) \tag{7}$$

$$\sigma(X) = h_\sigma(XW_\sigma + B_\sigma) \tag{8}$$

where h and B are activation functions and biases for NN layers respectively; $W_1 \in \mathbb{R}^{d \times c}$, $W_\mu \in \mathbb{R}^{(d+c) \times d'}$ and $W_\sigma \in \mathbb{R}^{d \times d'}$ are trainable weight matrices of MLP (6), expectation (7) and variance (8) for a NN layer respectively; $(m_i)_{i=1}^n = \mu(X, Y^*)$ is an output of (7) layer with $m_i \in \mathbb{R}^{d'}$; $concat(\cdot, \cdot)$ is a matrix concatenation column-wise.

To avoid the issues with high variance of the gradient estimation of $\mathbb{E}_{q_\phi(z|x,y^*)}\left[\log p_\theta(x|z, y^{pr})\right]$ by Monte Carlo method, we follow [3] in using the reparameterization trick to compute a low-variance gradient estimator for $q_\phi(z|x, y^*)$:

$$q_\phi(z|x, y^*) \sim Z, \quad Z = \mu(X, Y^*) + \sigma(X) \odot \epsilon, \epsilon \sim \mathcal{N}(0, I)$$

where $\odot$ is an element-wise product and ϵ is a random variable.

Now we can define (1) as a sequential sublayers in $PR(Y^*, \mu(Y^*, X), A)$:

1. the reweighting of A:

$$w_{i,j} = h(m_i, m_j); \ \forall \ w_{i,j} \in W; \tag{9}$$

$$A' = A + \gamma W; \tag{10}$$

where γ is a parameter of involvement W in reweighting of A within the range $[0, 1]$. Here we compute the similarities between the outputs of (7) because we assume that the expectation of the latent variable z more correctly defines the differences between nodes in latent space.

2. the regularization of A':

$$\hat{A}' = D'^{(-\sigma)} A' D'^{(\sigma-1)}; \ D'_{i,i} = \sum_{j=1}^n A'_{i,j} \tag{11}$$

where σ is a parameter for selection of regularization type: $\sigma = 1$ is a Standard Laplacian; $\sigma = 0$ is a PageRank; $\sigma = 1/2$ is a Normalized Laplacian;

3. the redefined PRSSL [1]:

$$F^t = \alpha \hat{A}' F^{t-1} + (1 - \alpha) Y^*; \ t \geq 0; \tag{12}$$

where $F^0 = Y^*$ (6) and F^t is a result of the t-th iterations, smoothly changing the node labels Y^* during iterations.

The final layer is the reconstruction of nodes (papers) $\hat{X} = f_\theta(Z, F^t)$ where $\hat{X} \in \mathbb{R}^{n \times d}$:

$$f_\theta(Z, F^t) = h_2(concat(Z, F^t)W_2 + B_2) \tag{13}$$

where $W_2 \in \mathbb{R}^{(d'+c) \times d}$, B_2 are weight and bias for x generation $p_\theta(x|z, y^{pr}) = f_\theta(z, y^{pr})$. We can turn to transductive regime of the aforementioned GenPR layers architecture by using modified loss as in (5). The Fig. 2 presents the difference between inductive (I) and transductive (T) GenPR architectures.

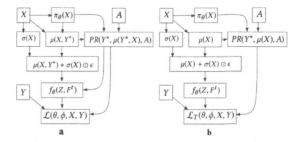

Fig. 2. The I-inductive (a) and T-transductive (b) architectures of GenPR.

4   Experimental Setup

For conducting an experiment in the graph-based SSL area we have taken following citation-graph data sets: Citeseer, Cora-ML, Pubmed, MS-Academic (MSA). The description and statistics of data sets available in work APPNP [6]. These data sets are available here[1].

As a baseline algorithms we have considered: the classical diffusion - based linear algorithms (LP [9], PRSSL [1]); the recent graph convolution - based NN (GCN [5], GAT [8], APPNP [6]); the deep generative model (M2 [4]). To avoid overfitting issue we applied L_2 regularization with parameter $\lambda = 0.05$ for weights $W_.$, dropout for $\hat{A}'$ with rate $dr = 0.5$ at each PowerIteration step and learning rate $l = 0.001$ for Adam optimizer. Moreover, we have used the random train-test-validation splitting strategy described in [6] and repeated experiments on each data set 500 times. For a fair model comparison we have made an architecture and parameters of GenPR that are very close to APPNP and GCN. In particular, for all data sets use the intermediate embedding layer $f_0(X) = relu(XW_0 + B_0)$ with $W_0 \in \mathbb{R}^{d \times \hat{d}}$ as the input for (6) with $\hat{d} = 64$, $W_1 \in \mathbb{R}^{\hat{d} \times c}$ and $h_1(\cdot) = softmax(\cdot)$, $d' = 64$ in (7) and (8), $\sigma = 0.5$ and $t = 4$ in (11), $B_. = 0$. In (12) for MSA $\alpha = 0.8$, for Cora-ML, Pubmed and Citeseer $\alpha = 0.9$.

We have selected the specific parameters of GenPR by the 5 fold cross-validation grid search[2]. For all data sets use $h(m_i, m_j) = (m_i^T m_j)^3$ in (9) and $\beta = 0.001$ in (4) and (5). In particular, we have used: for Citeseer: $\gamma = 1$ in (10), $h_\mu(\cdot) = h_\sigma(\cdot) = relu(\cdot)$ in (7) and (8), $h_2(\cdot) = sigmoid(\cdot)$ in (13); for Cora-ML, Pubmed and MSA: $\gamma = 0.001$ in (10), and $h_.(\cdot) = linear(\cdot)$.

5   Experimental Results

Table 1 presents performance of the classification based on the default adjacency matrix A or on the node features X leads to loss of classification quality because we do not use all available information. In the case of the combination of X

---

[1] https://github.com/klicperajo/ppnp/tree/master/ppnp/data.
[2] https://scikit-learn.org/stable/modules/grid_search.html.

Table 1. Average accuracy (%) on citation graphs. $\triangle$ and $\blacktriangle$ denote the statistical significance (t-test) of GenPR for $p < 0.05$ and $p < 0.01$, respectively, compared to the APPNP.

Input	Data set	Citeseer	Cora-ML	Pubmed	MSA
A	PRSSL	71.21	78.12	72.51	76.12
	LP	45.32	68.31	63.12	65.32
X	M2	70.81	79.22	77.6	86.12
X,A	APPNP	75.74	85.09	79.71	93.28
	GAT	75.43	84.41	77.73	91.18
	GCN	75.31	83.52	78.65	92.09
	GenPR (I)	**77.18**$^{\blacktriangle}$	85.52$^{\triangle}$	80.09$^{\triangle}$	**94.08**$^{\blacktriangle}$
	GenPR (T)	76.91$^{\triangle}$	**86.19**$^{\blacktriangle}$	**81.13**$^{\blacktriangle}$	93.81$^{\triangle}$

and A, GenPR significantly and consistently outperforms the others due to the intuition that default A contains incomplete information about nodes similarity. Since we have reached the best results with GenPR (I) and $\gamma = 1$ for Citeseer, it means that latent information is helpful for reweighting default adjacency matrix A (citation graph). In particular, Fig. 1 (c) shows that GenPR can be used for the explanation of classification results, by filter the edges by weight and observe nodes with more influence on considered one (e.g. node 545).

Figure 3 shows the GenPR outperforms the APPNP not only in terms of accuracy, but also in number of PowerIteration steps, because GenPR takes less steps to converge for better accuracy than APPNP. Moreover, the GenPR is less complex than APPNP because it uses just one layer for MLP rather than 2 in APPNP.

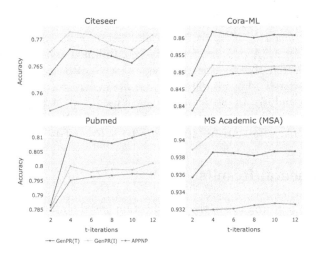

Fig. 3. Average accuracy of GenPR (I) inductive, GenPR (T) transductive and APPNP over the t-iteration steps.

6   Conclusion

In this work, we propose a graph-based SSL (I)/(T) framework created by embedding PRSSL in generative model. Based on the experimental results, we show that the generative model application for PRSSL can be used not only for the label spreading improvement, but also for interpretation of the classification results. We also show that GenPR significantly and consistently outperforms all other algorithms on every data set and requires less number of PageRank PowerIteration steps. Since GenPR produces complete weighted graph defined by A' we can use PageRank properties to split A' into batches that are complete subgraphs, which opens an opportunity to explore a distributed version of GenPR. The other interesting direction to investigate is an application of GenPR on data sets without default graph structure (e.g. images).

References

1. Avrachenkov, K., Mishenin, A., Gonçalves, P., Sokol, M.: Generalized optimization framework for graph-based semi-supervised learning. In: Proceedings of the 2012 SIAM International Conference on Data Mining, pp. 966–974. SIAM (2012)
2. Brezinski, C., Redivo-Zaglia, M.: The pagerank vector: properties, computation, approximation, and acceleration. SIAM J. Matrix Anal. Appl. **28**(2), 551–575 (2006)
3. Kingma, D.P., Welling, M.: Auto-encoding variational bayes. In: Proceedings of the 2nd International Conference on Learning Representations, ICLR (2013)
4. Kingma, D.P., Mohamed, S., Rezende, D.J., Welling, M.: Semi-supervised learning with deep generative models. In: Advances in neural information processing systems, pp. 3581–3589 (2014)
5. Kipf, T.N., Welling, M.: Semi-supervised classification with graph convolutional networks. In: 5th International Conference on Learning Representations, ICLR (2017)
6. Klicpera, J., Bojchevski, A., Günnemann, S.: Predict then propagate: graph neural networks meet personalized pagerank. In: International Conference on Learning Representations, ICLR (2019)
7. Page, L., Brin, S., Motwani, R., Winograd, T.: The pagerank citation ranking: Bringing order to the web. Technical report, Stanford InfoLab (1999)
8. Veličković, P., Cucurull, G., Casanova, A., Romero, A., Lio, P., Bengio, Y.: Graph attention networks. In: 6th International Conference on Learning Representations, ICLR (2018)
9. Zhu, X., Ghahramani, Z.: Learning from labeled and unlabeled data with label propagation. CMU Technical report (2002)

Finding New Multiword Expressions for Existing Thesaurus

Petr Rossyaykin[(✉)] and Natalia Loukachevitch

Lomonosov Moscow State University, Moscow 119991, Russian Federation
petrrossyaykin@gmail.com, louk_nat@mail.ru

Abstract. In this paper we study the task of adding new multiword expressions (MWEs) into an existing thesaurus. Standard methods of MWE discovery (statistical, context, distributional measures) can efficiently detect the most prominent MWEs. However, given a large number of MWEs already present in a lexical resource those methods fail to provide sufficient results in extracting unseen expressions. We show that the information deduced from the thesaurus itself is more useful than observed frequency and other corpus statistics in detecting less prominent expressions. Focusing on nominal bigrams (Adj-Noun and Noun-Noun) in Russian, we propose a number of measures making use of thesaurus statistics (e.g. the number of expressions with a given word present in the thesaurus), which significantly outperform standard methods based on corpus statistics or word embeddings.

Keywords: Multiword expressions · MWE · MWE extraction · Thesaurus · Lexical resources

1 Introduction

Multiword expressions (MWE) are recurring, conventionalized expressions that exhibit some kind of idiosyncrasy – semantic (e.g. *red tape*), morphosyntactic (e.g. *by and large*), statistical (e.g. *black and white*) [2]. Given the compositional nature of most computational methods and non-compositional syntactic and semantic properties of MWEs, they present a known difficulty for NLP [25,32,33]. However, automatic MWE identification is a prerequisite for semantically-oriented downstream applications. It was shown that MWE recognition can improve performance in machine translation [3,5], syntactic parsing [7], information retrieval [1], etc.

Most state-of-the-art systems of automatic MWE identification exhibit particularly strong sensitivity to unseen data. As was argued recently in [33] on the basis of the results of the PARSEME shared task, in order to make strong headway in MWE identification it should be coupled with MWE discovery, via syntactic lexicons.

The task of MWE discovery (or extraction) consists in extracting lists of potential MWEs from corpora and storing them for future use in an application-oriented resource. In this paper we experiment on MWE discovery for Russian

© Springer Nature Switzerland AG 2020
A. Filchenkov et al. (Eds.): AINL 2020, CCIS 1292, pp. 166–180, 2020.
https://doi.org/10.1007/978-3-030-59082-6_13

wordnet RuWordNet [21]. Previous studies using RuWordNet as a gold standard for MWE discovery reported very high average precision achieved with word embeddings [20] and clustering of measures [31]. In contrast to them and most other work on MWE extraction for lexical resources (see Sect. 2) our goal is to find unseen MWEs, i.e. expressions not yet included in the thesaurus. This task is especially relevant since new MWEs are continuously created as languages evolve [9,14].

Moreover, standard methods of MWE discovery rely exclusively or mostly on corpus statistics. In many cases the combination of syntactic information with raw observed frequency of the candidate expressions appears to be the most powerful discovery method [15,38]. However, MWEs' distribution in corpora follows Zipf's law: few MWE types occur frequently in texts, with a long tail of MWEs occurring rarely [39]. Thus, a lot of less prominent expressions remain undetected.

In this paper we show that the efficiency of standard methods significantly degrades with the most frequent expressions being already discovered. We present a novel method to supplement a thesaurus with less prominent but relevant MWEs using the information about single words and expressions already included in the thesaurus.

The remainder of this paper is structured as follows. In Sect. 2 we discuss previous work with particular focus on MWE discovery for supplementing thesauri and approaches making use of lexical resources. Section 3 presents the data (corpus, thesaurus, candidate expressions). Section 4 describes the measures based on thesaurus statistics proposed in this paper. Section 5 describes distributional (embedding-based) measures used for comparison. In Sect. 6 we present the results of our experiments and their discussion with the conclusion drawn in Sect. 7.

2   Related Work

Numerous methods proposed for MWE extraction make use of different types of idiosyncrasy. In general, those methods can be divided into the following groups: statistical, morphosyntactic, context, distributional and those based on external lexical resources (e.g., dictionaries, machine translation, parallel corpora, thesauri, etc.). We do not discuss measures based on morphosyntactic features, because we experimented on a lemmatized corpus. The measures from other groups were used for comparison with our method and are discussed below.

Pecina [27] compares 55 statistical association measures (AMs) as well as their combinations (obtained with the help of machine learning) on German and Czech data and shows that combinations outperform individual measures. Hoang et al. [16] compare 82 AMs on English Verb Particle Construction and Light Verb Constructions with standard AMs performing better than the ones using context information. Garcia et al. [15] run their experiments on English, Spanish and Portuguese data and achieve the best results with raw frequency and the measures promoting recurrent word combinations.

The use of context measures follows from the assumption that the immediate contexts in which candidate expressions appear can be informative for their detection. Riedl and Biemann [30] observe that MWEs tend to have contexts similar to single words and propose an efficient measure based on this property. Farahmand and Martins [13] achieve high precision in English MWE extraction combining a range of context-based features with frequency information.

Distributional measures use either count-based vectors [6,34] or word embeddings [12] to detect non-compositional phrases. Recently, contextualized word embeddings [8,28] were used for MWE discovery [22]. However, it was shown that, in contrast to many other tasks, "static" word embeddings (e.g., word2vec [23]) are superior to contextualized in detecting non-compositionality [25].

The measures proposed in this paper (see Sect. 4) take advantage of the information derived from a lexical resource and for this reason the last class of approaches is the most relevant for present discussion.

Similarly to this paper, Dubremetz and Nivre [10] focus on the extraction of nominal bigrams (Noun-Noun and Noun-Adj). They train several classifiers using statistical AMs as features and expressions found in the French Europarl corpus and present in the dictionary of French MWEs Delac [36] as instances of the positive class.

Piasecki et al. [29] use Polish wordnet (plWordNet) as a gold standard in the task of wordnet expansion. They utilize a number of AMs including the ones proposed in their paper to provide rankings of candidate expressions. The rankings are combined with different weights trained on a tuning corpus. It is shown that combinations of measures outperform individual AMs.

McCrae [22] focuses on the discovery of English multiword neologisms of Adj-Noun type. Expressions already present in WordNet [24] and expressions extracted randomly from Wikipedia and absent from WordNet were used to form the positive and negative classes respectively. The frequencies with which words appeared in positive and negative training sets were used to estimate probability of a word pair to be a neologism. Those probabilities, combined with both static and contextualized embeddings, formed vector representations of candidate expressions used in classification.

A wide range of papers deal with Russian MWE extraction and thesaurus extension. Zakharov [40] compares statstical AMs and combinations of ranks assigned by them in the task of extraction of MWEs containing selected nouns. Tutubalina and Braslavski [37] use learning-to-rank algorithms and a wide range of statistical, linguistic and metalinguistic features (e.g. match with a Wikipedia article title) for nominal MWE extraction. Braslavski and Sokolov [4] utilize context-based measures for discovery of biological terms of variable length. A number of methods make use of word2vec embeddings. Enikeeva and Mitrofanova [11] train linear transformation matrices to predict a collocate given a base word. Loukachevitch and Parkhomenko [20] use the proximity of candidates' vectors to the vectors of thesaurus expressions as a ranking measure.

Recently, several tasks similar to MWE extraction were put forward. Kopotev et al. [17] present a system for collocation and colligation extraction with the latter being defined as cooccurrence of word forms with grammatical phenomena. Kutuzov et al. [19] discuss the task of construction extraction. Construction refers to an MWE in which a variable can be replaced with another word form of the same semantic class, i.e. a set of collocations with the same base and different collocates. Clustering algorithms are compared in their ability to build clusters corresponding to constructions. Finally, a task of Taxonomy Enrichment for Russian (but for single words) was presented by Nikishina et al. [26]. The participants were asked to extend an existing taxonomy with previously unseen words: for each new word their systems should have provided a ranked list of possible (candidate) hypernyms present in RuWordNet.

In contrast to most of the work discussed here, we do not use thesaurus expressions to train classifiers with statistical AMs as features. Single words and MWEs present in RuWordNet are used to obtain thesaurus statistics (see Sect. 4). Measures based on this statistics are used to rank candidate expressions extracted from a corpus. Candidate expressions are not yet present in the thesaurus and are not seen during the computation of thesaurus statistics.

3   Data

3.1   Corpus, Thesaurus and Candidate Set

The corpus we experimented on was composed of news' texts from the Russian Internet published in 2011. It consists of 448 million tokens. After deleting all punctuation, lemmatizing and uppercasing it, we extracted all bigrams occurring not less than 50 times. We then used DeepPavlov[1] morphological tagger to filter out everything but Adj-Noun and Noun-Noun word pairs. We also deleted all expressions containing non-Cyrillic symbols or words shorter than 3 symbols to form the candidate set of 200 254 expressions (we refer to them as corpus or candidate expressions).

Since we define the task as supplementation of RuWordNet, the next step was to delete from candidate expressions MWEs already present there. The version of RuWordNet we used included 58 086 single words and 55 364 MWEs (not only bigrams), of which 15 794 were found in our candidate set and excluded from it. This resulted in the candidate set with the size of 184 460 expressions.

3.2   Candidate Set Annotation

Expressions suggested by human experts as eligible for further inclusion in RuWordNet were recognized as potential members of positive class [26]. We PoS-tagged and lemmatized them with the help of DeepPavlov and pymystem[2] respectively. 3 711 of those expressions were found in our candidate set and thus formed the positive class. The remaining expressions formed the negative class.

---

[1] https://docs.deeppavlov.ai/en/0.0.8/components/morphotagger.html.
[2] https://github.com/nlpub/pymystem3.

Such a small amount of positive candidates is due to the fact that most MWEs are already described in RuWordNet.

Since some true MWEs present in our corpus could still be unseen by the human experts having proposed supplementations for RuWordNet, we performed additional annotation. We combined top-100 expressions extracted by each individual measure (Sects. 4 and 5) into a single list and annotated them manually. Those 173 expressions which were considered true MWEs by each author were included in the positive class. This resulted in the candidate set consisting of 3 884 positive (MWE) and 180 576 negative expressions respectively. The description of data is summarized in Fig. 1.

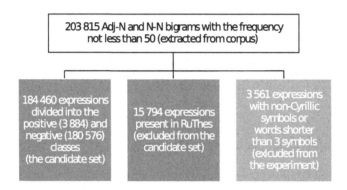

Fig. 1. The set of candidate expressions used in the experiment.

4   Thesaurus-Based Measures

We experimented with three measures based on thesaurus statistics, which we call TfCf (short for Thesaurus frequency/Corpus frequency), TpCf (Thesaurus probability/Corpus frequency) and TpDev (Thesaurus probability * Deviation). Each of them is defined relatively to a single word, rather than an entire bigram. As a result, for a given bigram '$w_1\_w_2$', five variants of each measure are computed. They are defined below exemplified by TfCf:

$$TfCf1(w_1\_w_2) = TfCf(w_1) \tag{1}$$

$$TfCf2(w_1\_w_2) = TfCf(w_2) \tag{2}$$

$$TfCf\text{-}max(w_1\_w_2) = max(TfCf1, TfCf2) \tag{3}$$

$$TfCf\text{-}min(w_1\_w_2) = min(TfCf1, TfCf2) \tag{4}$$

$$TfCf\text{-}mean(w_1\_w_2) = \frac{(TfCf1 + TfCf2)}{2} \tag{5}$$

We proceed now to the description of TfCf, TpCf and TpDev. The first measure (or rather a group of measures) TfCf is based on the following assumption: if a word appears in numerous candidate expressions it is uninformative and thus unlikely to form MWEs. On the other hand, if a word appears in many thesaurus expressions, it is highly informative and is likely to be a part of unseen MWEs. With this, TfCf is defined as follows (Tf and Cf are short for Thesaurus frequency and Corpus frequency, respectively):

$$TfCf(w) = \frac{Tf(w)}{Cf(w)} \tag{6}$$

Where Tf(w) is the number of occurrences of a given word w in thesaurus expressions (55 364 RuWordNet expressions in our case) plus smoothing value 1 and Cf(w) is the number of occurrences of w in corpus expressions (types, not tokens; 184 460 expressions in our case) plus 1. We added 1 to both numbers since for many words from corpus expressions TfCf could turn into 0 due to their absence in thesaurus expressions. We also experimented on TfCf variants with 0.1 and 0.01 instead of 1 as smoothing values for Tf and Cf (those variants are called TfCf-10 and TfCf-100 respectively).

The next measure Tp is defined as follows:

$$Tp(Tf(w)) = \begin{cases} \frac{n(Tf(w)+1)}{n(Tf(w))}, & n(Tf(w)+1) > 0. \\ Tp(Tf(w)-1), & n(Tf(w)+1) = 0. \end{cases} \tag{7}$$

Similar to (6), Tf(w) is the number of occurrences of w in thesaurus expressions, n(x) is cardinality of the set of words with Tf = x. If, for instance, there are 10 words with 40 occurrences in thesaurus expressions and 2 words with 41 occurrences, then n(40) = 10 and n(41) = 2. Then, for a word w with 40 occurrences in thesaurus expressions Tp(Tf(w)) = Tp(40) = 2/10 = 0.2. We assume that low Tp(Tf(w)) indicates that w is unlikely to obtain a new expression containing it.

Figure 2 provides the plot showing the values of Tp(Tf)*100 and n(Tf) for Tf from 5 (words with 5 expressions in RuWordNet) to 60. As can be seen, there are 598 words which appear in RuWordNet expressions 5 times. However, n(Tf) decreases drastically. E.g. there are only 44 words having 20 thesaurus expressions. With the decrease of n(Tf) the fluctuation of Tp(Tf)*100 increases.

Average precision of raw Tp turned out to be relatively small in our experiments, so in Sect. 6 we provide the results only for the following variant:

$$TpCf(w) = \frac{Tp(Tf(w))}{Cf(w)} \tag{8}$$

Another measure combining Tp and Cf makes use of the median of Cf. In our case median(Cf) = 2. The idea is to penalize words which occur in corpus expressions too often or too rarely. TpDev is defined in (10):

$$Dev(w) = \begin{cases} \frac{median(Cf)}{Cf(w)^2}, & Cf(w) \geq median(Cf). \\ \frac{Cf(w)}{median(Cf)^2}, & Cf(w) < median(Cf). \end{cases} \tag{9}$$

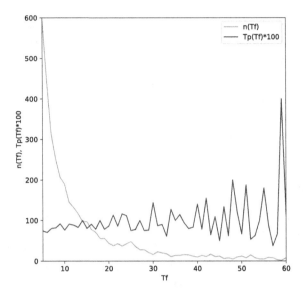

Fig. 2. n(Tf) and Tp(Tf)*100 for Tf from 5 to 60.

$$TpDev(w) = Tp(w) * Dev(w) \tag{10}$$

5   Statistical, Context and Distributional Measures

For the purposes of comparison we computed the same 30 statistical and context
AMs as in [31]. They are presented in Table 6 (see Appendix). The results for
the best of them are reported in Sect. 6. We also experimented on a number of
distributional measures which will be discussed in this section in more detail.
Both static (word2vec) and contextual (RuBERT [18]) word embeddings were
used.

Word2vec Embeddings. We used word2vec model with the following parameters: vector size 200, window size 3, min_count 3 (other parameters left default).
We concatenated candidate bigrams into single tokens with underscore ('x y'
→ 'x_y') in order to build embeddings for them. It was shown in [35] that this
method yields low quality embeddings for rare noun compounds, however, since
we do not consider expressions with the observed frequency lower than 50, this
problem is not relevant. We use the following measures to estimate the non-
compositionality of a given candidate bigram '$w_1\_w_2$':

$$left(w_1, w_2) = cos(v(w_1\_w_2), v(w_1)) \tag{11}$$

$$right(w_1, w_2) = cos(v(w_1\_w_2), v(w_2)) \tag{12}$$

$$mean(w_1, w_2) = \frac{left(w_1, w_2) + right(w_1, w_2)}{2} \qquad (13)$$

$$centroid(w_1, w_2) = cos(v(w_1\_w_2), \frac{v(w_1) + v(w_2)}{2}) \qquad (14)$$

Where v(x) is the embedding of x, cos(x, y) is cosine similarity between x and y.

We also experimented on distributional measures DFsing and DFthes proposed in [20], as well as our modification DFcol. DFsing is the similarity between the phrase embedding $v(w_1\_w_2)$ and the embedding of the most (cosine) similar single word w which should be different from the phrase components:

$$DFsing(w_1, w_2) = max(cos(v(w_1\_w_2), v(w))) \qquad (15)$$

where w is a word from the model vocabulary distinct from w_1 and w_2.

DFthes is the similarity between the phrase embedding $(w_1\_w_2)$ and the embedding of the most similar thesaurus entry (having an embedding). Finally, DFcol is DFthes restricted to thesaurus bigrams (observed in the corpus but excluded from the candidate set, see Sect. 3.1).

$$DFthes(w_1, w_2) = max(cos(v(w_1\_w_2), v(w_{thes}))) \qquad (16)$$

$$DFcol(w_1, w_2) = max(cos(v(w_1\_w_2), v(x\_y_{thes}))) \qquad (17)$$

where w_{thes} is a thesaurus entry (either single word or bigram), x_y_{thes} is a thesaurus bigram.

RuBERT Embeddings. We also computed the measures in (11–14) using BERT (RuBERT in our case). We tried out 3 methods of building static embeddings of words and expressions out of contextualized ones outputted by BERT. In the following discussion contextualized embedding refers to the hidden states at the output of the last hidden layer of BERT.

1. Approximating across contexts. BERT tokenizes every input sentence into vocabulary items (words and morphs), which have vocabulary vector representations and can be fed into the model. We tokenize each candidate expression to form the set of vocabulary items appearing in candidate expressions. For each vocabulary item we build a static embedding by summing its contextualized embeddings across 20 contexts from the corpus. We do not approximate across all the contexts because (as a result of RuBERT being pre-trained) even a single contextualized embedding is meaningful at the type level. For each word or bigram, its static embedding is then the sum of static embeddings of morphs appearing in its tokenized form.
2. [SEP] token embedding. We regard a word/bigram x as a sentence and use the contextualized embedding of sentence-final token [SEP] as the static embedding of x.
3. Sum of token embeddings. We regard a word/bigram x as a sentence and use the sum of contextualized token embeddings outputted by the model as the static embedding of x.

6   Results

The measures presented in 4 and 5 were used to rank the candidate expressions. To evaluate the list rankings, we utilized uninterpolated average precision measure (AP). Table 1 presents the results for the best statistical and context AMs. Measures promoting frequent expressions turned out to be the most powerful ones. They provide very little improvement in comparison to raw observed frequency and average precision is generally poor across AMs (note, however, that the positive class amounts to only 2.11% of the candidate expressions).

Table 1. Average precision of the best statistical and context AMs

	AP@50	AP@100	AP@500	AP@1000
Frequency	0.100	0.075	0.039	0.032
LLR	**0.108**	**0.083**	0.05	0.043
Poisson	0.107	0.083	0.05	0.043
True PMI	0.106	0.082	0.05	0.041
freq-CI	0.093	0.073	**0.06**	**0.052**

Table 2 presents the results for the best distributional measures. Measures using thesaurus information provide significant improvement in comparison to those based exclusively on corpus data (*left, right* and *DFsing* among the presented in Table 2). As far as BERT embeddings are concerned, static embeddings of the second and the third types (which we call SEP and sum respectively) appeared to capture non-compositionality better that those based on approximating across contexts. BERT embeddings also outperform word2vec ones (without the usage of thesaurus information) which is quite surprising given that our task required non-contextualized word representations.

Table 2. Average precision of the best distributional AMs

	AP@50	AP@100	AP@500	AP@1000
word2vec				
Left	0.075	0.068	0.043	0.036
DFsing	0.034	0.034	0.028	0.025
DFthes	0.307	0.234	0.105	0.074
DFcol	**0.35**	**0.254**	**0.111**	**0.079**
RuBERT				
right-SEP	0.097	0.077	0.035	0.030
right-sum	**0.153**	**0.121**	**0.062**	**0.052**

Table 3 presents the results for the best thesaurus-based measures. In this case, in contrast to distributional measures, the information from both components of a candidate expression is meaningful (we observe only *min* and *mean* variants among the best measures). The most complex measures TpCf and TpDev turned out to be the best among the thesaurus based measures and in general.

Table 3. Average precision of the best thesaurus-based AMs

	AP@50	AP@100	AP@500	AP@1000
TfCf-min	0.256	0.231	0.157	**0.13**
TfCf-min-100	0.267	0.24	0.155	0.124
TpCf-mean	0.376	**0.311**	**0.162**	0.12
TpDev-min	**0.392**	0.271	0.101	0.076

The comparison of the best measures across all 3 groups is provided in Table 4 and on Fig. 3. As was noted above, standard AMs showed poor performance given the absence of the most frequent MWEs among candidates expressions. The best statistical AM (LLR) is considerably outperformed by the best measures from other groups, except for BERT. BERT outperforms LLR slightly and this result looks promising given rather simplistic ways to build static embeddings we used. Note, however, that BERT results varied very strongly. We report the best results, but for many combinations of measure and the way of static embedding generation AP was very low.

Next, word2vec-based measure DFcol which ranks candidates according to their cosine proximity to the embeddings of RuWordNet expressions shows that the usage of thesaurus information is indeed useful for the discovery of less prominent MWEs. However, it is significantly inferior to the best of the thesaurus-based measures proposed in this work.

Table 4. Average precision of the best measures across all 3 groups

	AP@50	AP@100	AP@500	AP@1000
LLR	0.108	0.083	0.05	0.043
DFcol	0.35	0.254	0.111	0.079
right-sum	0.153	0.121	0.062	0.052
TfCf-min-100	0.267	0.24	0.155	**0.124**
TpCf-mean	0.376	**0.311**	**0.162**	0.12
TpDev-Min	**0.392**	0.271	0.101	0.076

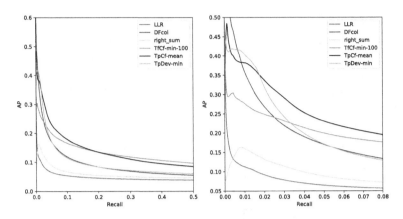

Fig. 3. AP of the best individual measures with the recall up to 0.5 and 0.08

In Table 5 we provide the lists of top-10 expressions extracted by LLR and TpCf-mean. Note that LLR promoted proper names and other recurrent but not relevant expression. TpCf by virtue of its definition is immune to this. In general, as the table witnesses, TpCf-mean promoted on top much more relevant expressions.

Table 5. Top-10 expressions extracted by LLR and TpCf-mean

№	LLR expression	cls	TpCf-mean expression	cls
1	РИА Новости 'RIA News'	N	шерстяной носок 'woolen sock'	N
2	настоящее время 'present time'	T	бутылочное горлышко 'bottle neck'	N
3	Дмитрий Медведев 'Dmitriy Medvedev'	N	головоногий моллюск 'cephalopod'	T
4	такий образ 'such a way'	N	сушеный гриб 'dried mushroom'	N
5	прошлый год 'last year'	N	бельевая верёвка 'clothesline'	T
6	прошлое год 'past year'	N	спиртовая настойка 'alcoholate'	T
7	Владимир Путин 'Vladimir Putin'	N	плодовая мушка 'drosophila'	T
8	млн рубль 'mln ruble'	N	соломенная шляпа 'straw hat'	N
9	Единая Россия 'United Russia'	T	конский хвост 'horsetail'	T
10	внутреннее дело 'internal affair'	N	телец близнецы 'Taurus Gemini'	N

7   Conclusion

We presented an efficient method to supplement a thesaurus with new multiword expressions. Our experiments were based on RuWordNet, a large lexical resource of Russian, already containing more than 50 000 MWEs. Both statistical AMs and word embeddings rely exclusively on the distribution of words in corpora to discover potential MWEs. It turns out that the significance of this information degrades substantially with the most prominent expressions being already extracted. Thus, there is a long tail of MWEs which remain undiscovered.

On the other hand, as we showed, the distribution of the words and expressions in a thesaurus is meaningful and can be used for discovery of unseen MWEs relevant for this resource. Combined with corpus statistics, this information allows to achieve high precision of MWE extraction. By virtue of this, we proposed a number of novel measures, which outperformed standard unsupervised methods drastically.

We experimented on Russian nominal bigrams, but the proposed measures are basically language- and resource-independent and can be adopted in future work to other languages and classes of MWEs. Crucially, given their difference from statistical AMs and context measures, they can also be used as informative features in machine learning, which we leave for future work.

Acknowledgments. Current study is supported by the RFBR foundation (project N 18-00-01226 (18-00-01240)). We would like to thank the three anonymous AINL reviewers for valuable comments and advice.

Appendix

In Table 6 N is the number of tokens in corpus, xy is a bigram consisting of words x and y, $f(x)$ is the observed frequency of x, $P(x) = f(x)/N$, $\overline{x}$ stands for any word except x, $r(x)$ is a set of unique words occurring in corpus immediately to the right from x, $l(x)$ is a set of unique words occuring in corpus immediately to the left from x, $\sum_{x,y} A(x,y) = A(x,y) + A(\overline{x},y) + A(x,\overline{y}) + A(\overline{x},\overline{y})$, W stands for any word which does not form a candidate expression with an adjacent word x in xW or y in Wy

Table 6. Statistical association and context measures used in the experiment

Name	Formula	Name	Formula
Frequency	$f(xy)$	Normalized PMI	$\frac{PMI(xy)}{-log(P(xy))}$
PMI	$log\frac{P(xy)}{P(x)P(y)}$	Normalized MI	$\frac{\sum_{x,y}P(xy)log\frac{P(xy)}{P(x)P(y)}}{-\sum_{x,y}P(xy)*log(P(xy))}$
Dice coefficient	$\frac{2f(xy)}{f(x)+f(y)}$	MI/NF(0.5)	$\frac{MI}{0.5P(x)+0.5P(y)}$
LLR	$2\sum_{x,y}f(xy)log\frac{P(xy)}{P(x)P(y)}$	PMI/NF(0.77)	$\frac{PMI}{0.77P(x)+0.23P(y)}$
chi-square	$\frac{(f(xy)-\frac{f(x)f(y)}{N})^2}{f(x)f(y)}$	MI/NF$_{max}$	$\frac{MI}{max(P(x),P(y))}$
Piatersky-Shapiro	$P(xy)-P(x)P(y)$	PMI/NF$_{max}$	$\frac{PMI}{max(P(x),P(y))}$
t-score	$\frac{f(xy)-\frac{f(x)f(y)}{N}}{\sqrt{f(xy)}}$	$NPMI_C$	$\frac{PMI(xy)}{-log(P(x))}$
Geometric mean	$\frac{f(xy)}{\sqrt{f(x)f(y)}}$	gravity count (GC)	$log\frac{f(xy)|r(x)|}{f(x)}+$ $+log\frac{f(xy)|r(y)|}{f(y)}$
Odds ratio	$log\frac{(f(xy)+0.5)(f(\overline{xy})+0.5)}{(f(\overline{x}y)+0.5)(f(x\overline{y})+0.5)}$	Modified GC	$log(\frac{f(xy)|r(x)|}{f(x)}+$ $+\frac{f(xy)|l(y)|}{f(y)})$
Poisson significance ratio	$(\frac{f(x)f(y)}{N}-$ $f(xy)log(\frac{f(x)f(y)}{N})++$ $log(f(xy)!))/log(N)$	type-LR	$\sqrt{|r(x)|*|l(y)|}$
Modified Dice	$log(f(xy))*DC(xy)$	type-FLR	$\frac{f(xy)}{type\text{-}LR}$
Confidence	$max(P(x|y),P(y|x))$	context intersection (CI)	$\frac{|l(xy)l(x)|}{|l(x)|}\times$ $\times\frac{r(xy)\cap r(y)}{|r(y)|}$
Local PMI	$f(xy)*PMI(xy)$	independent CI (ICI)	$\frac{|l(xy)\cap l(xW)|}{|l(xW)|}\times$ $\times\frac{r(xy)\cap r(Wy)}{|r(Wy)|}$
Augmented PMI	$log\frac{P(xy)}{P(\overline{x}y)P(x\overline{y})}$	CI-freq	$f(xy)*CI(xy)$
Cubic PMI	$log\frac{N*f(xy)^3}{f(x)f(y)}$	ICI-log(freq)	$log(f(xy))*ICI(xy)$

References

1. Acosta, O., Villavicencio, A., Moreira, V.: Identification and treatment of multi-word expressions applied to information retrieval. In: Proceedings of the Workshop on Multiword Expressions: From Parsing and Generation to the Real World, pp. 101–109 (2011)
2. Baldwin, T., Kim, S.N.: Multiword expressions. In: Handbook of natural language processing, vol. 2, pp. 267–292 (2010)
3. Bouamor, D., Semmar, N., Zweigenbaum, P.: Identifying bilingual multi-word expressions for statistical machine translation. In: LREC, pp. 674–679 (2012)
4. Braslavski, P., Sokolov, E.: Comparison of five methods for variable length term extraction. In: Computational Linguistics and Intellectual Technologies: Papers from the Annual conference "Dialogue", no. 7, pp. 67–74 (2008)
5. Carpuat, M., Diab, M.: Task-based evaluation of multiword expressions: a pilot study in statistical machine translation. In: Human Language Technologies: The 2010 Annual Conference of the North American Chapter of the Association for Computational Linguistics, pp. 242–245 (2010)
6. Van de Cruys, T., Moirón, B.V.: Semantics-based multiword expression extraction. In: Proceedings of the Workshop on A Broader Perspective on Multiword Expressions, pp. 25–32 (2007)

7. De Lhoneux, M.: Ccg parsing and multiword expressions. arXiv preprint arXiv:1505.04420 (2015)
8. Devlin, J., Chang, M.W., Lee, K., Toutanova, K.: Bert: Pre-training of deep bidirectional transformers for language understanding. arXiv preprint arXiv:1810.04805 (2018)
9. Dhar, P., van der Plas, L.: Learning to predict novel noun-noun compounds. arXiv preprint arXiv:1906.03634 (2019)
10. Dubremetz, M., Nivre, J.: Extraction of nominal multiword expressions in french. In: Proceedings of the 10th Workshop on Multiword Expressions (MWE), pp. 72–76 (2014)
11. Enikeeva, E.V., Mitrofanova, O.A.: Russian collocation extraction based on word embeddings. In: Computational Linguistics and Intellectual Technologies: papers from the Annual conference "Dialogue", pp. 52–64 (2017)
12. Farahmand, M., Henderson, J.: Modeling the non-substitutability of multiword expressions with distributional semantics and a log-linear model. In: Proceedings of the 12th Workshop on Multiword Expressions, pp. 61–66 (2016)
13. Farahmand, M., Martins, R.T.: A supervised model for extraction of multiword expressions, based on statistical context features. In: Proceedings of the 10th Workshop on Multiword Expressions (MWE), pp. 10–16 (2014)
14. Fazly, A., Cook, P., Stevenson, S.: Unsupervised type and token identification of idiomatic expressions. Comput. Linguist. $35(1)$, 61–103 (2009)
15. Garcia, M., Salido, M.G., Ramos, M.A.: A comparison of statistical association measures for identifying dependency-based collocations in various languages. In: Proceedings of the Joint Workshop on Multiword Expressions and WordNet (MWE-WN 2019), pp. 49–59 (2019)
16. Hoang, H.H., Kim, S.N., Kan, M.Y.: A re-examination of lexical association measures. In: Proceedings of the Workshop on Multiword Expressions: Identification, Interpretation, Disambiguation and Applications (MWE 2009), pp. 31–39 (2009)
17. Kopotev, M., Escoter, L., Kormacheva, D., Pierce, M., Pivovarova, L., Yangarber, R.: Online extraction of russian multiword expressions. In: The 5th Workshop on Balto-Slavic Natural Language Processing, pp. 43–45 (2015)
18. Kuratov, Y., Arkhipov, M.: Adaptation of deep bidirectional multilingual transformers for russian language. arXiv preprint arXiv:1905.07213 (2019)
19. Kutuzov, A., Kuzmenko, E., Pivovarova, L.: Clustering of Russian adjective-noun constructions using word embeddings. In: Proceedings of the 6th Workshop on Balto-Slavic Natural Language Processing, pp. 3–13 (2017)
20. Loukachevitch, N., Parkhomenko, E.: Recognition of multiword expressions using word embeddings. In: Kuznetsov, S.O., Osipov, G.S., Stefanuk, V.L. (eds.) RCAI 2018. CCIS, vol. 934, pp. 112–124. Springer, Cham (2018). https://doi.org/10.1007/978-3-030-00617-4_11
21. Loukachevitch, N.V., Lashevich, G., Gerasimova, A.A., Ivanov, V.V., Dobrov, B.V.: Creating Russian wordnet by conversion. In: Computational Linguistics and Intellectual Technologies: Papers From the Annual conference "Dialogue", pp. 405–415 (2016)
22. McCrae, J.P.: Identification of adjective-noun neologisms using pretrained language models. In: Proceedings of the Joint Workshop on Multiword Expressions and WordNet (MWE-WN 2019), pp. 135–141 (2019)
23. Mikolov, T., Sutskever, I., Chen, K., Corrado, G.S., Dean, J.: Distributed representations of words and phrases and their compositionality. In: Advances in neural information processing systems, pp. 3111–3119 (2013)

24. Miller, G.A.: Wordnet: a lexical database for English. Commun. ACM **38**(11), 39–41 (1995)
25. Nandakumar, N., Baldwin, T., Salehi, B.: How well do embedding models capture non-compositionality? a view from multiword expressions. In: Proceedings of the 3rd Workshop on Evaluating Vector Space Representations for NLP, pp. 27–34 (2019)
26. Nikishina, I., Logacheva, V., Panchenko, A., Loukachevitch, N.: Russe'2020: Findings of the first taxonomy enrichment task for the russian language. arXiv preprint arXiv:2005.11176 (2020)
27. Pecina, P.: A machine learning approach to multiword expression extraction. In: Proceedings of the LREC Workshop Towards a Shared Task for Multiword Expressions (MWE 2008), vol. 2008, pp. 54–61. Citeseer (2008)
28. Peters, M.E., et al.: Deep contextualized word representations. arXiv preprint arXiv:1802.05365 (2018)
29. Piasecki, M., Wendelberger, M., Maziarz, M.: Extraction of the multi-word lexical units in the perspective of the wordnet expansion. In: Proceedings of the International Conference Recent Advances in Natural Language Processing, pp. 512–520 (2015)
30. Riedl, M., Biemann, C.: A single word is not enough: ranking multiword expressions using distributional semantics. In: Proceedings of the 2015 conference on empirical methods in natural language processing, pp. 2430–2440 (2015)
31. Rossyaykin, P.O., Loukachevitch, N.V.: Measure clustering approach to MWE extraction. In: Computational Linguistics and Intellectual Technologies: Papers From the Annual conference "Dialogue", pp. 562–575 (2019)
32. Sag, I.A., Baldwin, T., Bond, F., Copestake, A., Flickinger, D.: Multiword expressions: a pain in the neck for NLP. In: Gelbukh, A. (ed.) CICLing 2002. LNCS, vol. 2276, pp. 1–15. Springer, Heidelberg (2002). https://doi.org/10.1007/3-540-45715-1_1
33. Savary, A., Cordeiro, S., Ramisch, C.: Without lexicons, multiword expression identification will never fly: A position statement. In: Proceedings of the Joint Workshop on Multiword Expressions and WordNet (MWE-WN 2019), pp. 79–91 (2019)
34. Senaldi, M.S.G., Lebani, G.E., Lenci, A.: Lexical variability and compositionality: Investigating idiomaticity with distributional semantic models. In: Proceedings of the 12th workshop on multiword expressions, pp. 21–31 (2016)
35. Shwartz, V.: A systematic comparison of english noun compound representations. arXiv preprint arXiv:1906.04772 (2019)
36. Silberztein, M.: Le dictionnaire delac. Dictionnaires électroniques du français, pp. 73–83 (1990)
37. Tutubalina, E., Braslavski, P.: Multiple features for multiword extraction: a learning-to-rank approach. In: Computational Linguistics and Intellectual Technologies. In: Proceedings of the Annual International Conference "Dialogue" (2016)
38. Wermter, J., Hahn, U.: You can't beat frequency (unless you use linguistic knowledge)-a qualitative evaluation of association measures for collocation and term extraction. In: Proceedings of the 21st International Conference on Computational Linguistics and 44th Annual Meeting of the Association for Computational Linguistics, pp. 785–792 (2006)
39. Williams, J., Lessard, P.R., Desu, S., Clark, E.M., Bagrow, J.P., Danforth, C.M., Dodds, P.S.: Zipf's law holds for phrases, not words. Sci. Rep. **5**, 12209 (2015)
40. Zakharov, V.P.: Automatic collocation extraction: association measures evaluation and integration. In: Computational Linguistics and Intellectual Technologies: Papers From the Annual conference "Dialogue", pp. 387–398 (2017)

Matching LIWC with Russian Thesauri: An Exploratory Study

Polina Panicheva[1,2]([⊠]) [iD] and Tatiana Litvinova[2] [iD]

[1] National Research University Higher School of Economics,
16 Soyuza Pechatnikov Street, St. Petersburg 190121, Russia
ppanicheva@hse.ru
[2] RusProfiling Lab, Voronezh State Pedagogical University,
86 Lenina Street, Voronezh 394043, Russia
centr_rus_yaz@mail.ru

Abstract. In Author Profiling research, there is a growing interest in lexical resources providing various psychologically meaningful word categories. One of such instruments is Linguistic Inquiry and Word Count, which was compiled manually in English and translated into many other languages. We argue that the resource contains a lot of subjectivity, which is further increased in the translation process. As a result, the translated lexical resource is not linguistically transparent. In order to address this issue, we translate the resource from English to Russian semi-automatically, analyze the translation in terms of agreement and match the resulting translation with two Russian linguistic thesauri. One of the thesauri is chosen as a better match for the psychologically meaningful categories in question. We further apply the linguistic thesaurus to analyze the psychologically meaningful word categories in two Author Profiling tasks based on Russian texts. Our results indicate that linguistically-motivated thesauri not only provide objective and linguistically motivated content, but also result in significant correlates of certain psychological states, replicating evidence obtained with hand-crafted lexical resources.

Keywords: Author Profiling · Lexical resources · Text-based personality prediction · Russian dictionary · Linguistic Inquiry and Word Count

1 Introduction

In recent years, the abundance of texts which people write, both online and offline, has led researchers to studying these texts as containing natural information on human behavior. In addition to traditional self-report approaches in psychology, a variety of personality characteristics and mental states can be analyzed by looking at the natural language traces of human behavior. To make such studies feasible, there is a strong need in a tool that would represent psychologically relevant categories linguistically, i.e., in a vocabulary. One of the most popular tools of this type, Linguistics Inquiry and Word Count (LIWC), was built in 1990-s by J. Pennebaker and colleagues for analysis of English texts [1–3].. Since then, LIWC has been translated and effectively used in many languages other than English [4–8]. However, in some languages which are

© Springer Nature Switzerland AG 2020
A. Filchenkov et al. (Eds.): AINL 2020, CCIS 1292, pp. 181–195, 2020.
https://doi.org/10.1007/978-3-030-59082-6_14

structurally very different from English (Arabic, Turkish and Russian), LIWC adaptations have never been empirically validated [6], which forced researchers to add additional manually constructed dictionaries [9, 10].

It is clear that when translating LIWC into other languages, a number of difficulties arise because of structural differences between languages and cultural differences represented therein. A natural way of overcoming these is to apply semantic information coded in lexical resources of the target language.

Moreover, there are more fundamental issues regarding LIWC construction and usage from the linguistics viewpoint. First, LIWC construction primarily employs expertise in psychology [11]. In the psychological community there is a high need for a ready-made text analytics tool, and a lack of resources for employing state-of-the-art NLP solutions. As a result, LIWC is widely used for psychological tasks. However, from the linguistic viewpoint, LIWC lexicon structure is not well justified. The second issue regards LIWC translation specifically: it is a time- and resource-consuming manual process, and the result is highly prone to subjectivity. A preliminary attempt to expand machine-translated LIWC categories in Russian (described in Sect. 3.1) resulted in extremely low agreement metrics.

A natural solution to these considerations is constructing a dictionary based on available linguistic resources in Russian. We set out to construct a preliminary Russian LIWC lexicon by matching existing Russian thesauri to a translated version of LIWC. We choose the following word categories from LIWC: **Cognitive, Social, Biological processes, Perceptual processes** and the contents thereof: **See, Feel, Hear**.

To our knowledge, this is the first attempt to explore the contents of LIWC dictionary in linguistic terms and expand the suggested categories with available lexical resources, either in Russian or in English.

We approach the task of matching LIWC with Russian thesauri in the following steps:

- We first perform machine translation of the LIWC 2015 vocabulary to Russian.
- The translated categories **Cognitive** and **Social** are further enriched by 3 annotators.
- The resulting word lists are matched with existing Russian thesauri.
- A single thesaurus is chosen based on the matching statistics. Every LIWC category in question is represented by a number of thesaurus categories.
- The resulting category structure is preliminary applied to Author Profiling tasks, namely in identifying personality and depression/anxiety lexical correlates.

By performing these steps, we answer the following research questions:

1. How are LIWC categories mapped onto existing Russian thesauri?
2. Is it possible to construct a preliminary LIWC-like dictionary by preserving an original Russian thesaurus category structure?
3. To what extent will the preliminary dictionary represent the original LIWC translations?
4. Can the preliminary dictionary be used as a starting point for author profiling tasks?

The rest of the paper is organized as follows. In Sect. 2, we review related work on LIWC development and translation into other languages; we describe the Russian thesauri available for our task and give a brief overview of LIWC application to Author

Profiling. Section 3 describes LIWC translation and matching with the selected thesauri in Russian, and the preliminary version of the Russian LIWC expanded with a thesaurus. In Sect. 4, we present experiments on Author Profiling in Russian with the preliminary thesaurus. In Sect. 5 we give our conclusions with regard to the stated research questions, and outline directions for future work.

2 Related Work

2.1 LIWC Construction and Translations

The original LIWC in English was first compiled in 1993, and the latest version was released in 2015 [11]. The original idea behind LIWC is to compile a dictionary of words denoting basic cognitive and emotional dimensions, which are often studied in sociology and psychology. The latest version includes around 90 variables, covering 6,400 words (more specifically, word stems). The LIWC dictionary is organized hierarchically and contains 21 linguistic categories (for example, parts of speech, function words per category, words longer than 6 letters, etc.), and 41 categories tapping psychological constructs (affective, social, cognitive processes, drives, informal language, etc.; extensive description can be found in [11]). The dictionary has been composed following an elaborate manual procedure involving 2–8 judges and recourse to common emotion rating scales and English dictionaries.

The 2007-version dictionary was translated into Russian [4]. The Russian dictionary includes 61 categories and around 5,900 word stems organized in a flat category structure. The Russian LIWC dictionary has never been validated for internal consistency or external validity. As a result, Russian LIWC was used in very few studies [9, 10] and required significant manual additions.

The first study accounts for the changes in various word category occurrences in a period prior to the suicide committed by an author of two weblogs [9]. Before application of the Russian LIWC dictionary, it was manually filtered, and a considerable number of Russian word categories were manually added. Across both weblogs, the occurrence of personal and overall pronouns, including *I*, drop, as the suicide date approaches. Moreover, the occurrence of words related to social processes drops immediately prior to the suicide date, indicating reduced social engagement in suicidal ideation. The second study identifies a number of stable LIWC categories in Russian texts by the same author across topics and truth/deception mode [10]. LIWC categories were filtered based on a minimum occurrence threshold. The highly stable categories include words related to space and cognitive processes. The moderately stable categories include personal and overall pronouns, emotion and negation words, words related to time. The results highlight the importance of pronoun categories for language-based personality profiling which is in line with numerous research in English [2, 12], as well as the importance of categories denoting cognitive and emotional dimensions.

Meanwhile, there have been a number of successful LIWC adaptations to other languages in recent years based on machine translation [7, 8]. The original LIWC category structure was mostly maintained, with a few language-specific changes, especially regarding function words categories. However, despite the fact that German and Dutch languages are closely related to English, an additional step of elaborate manual revision, including category refinement and expansion, was required to obtain high equivalence between the original and translated versions.

The translated dictionaries are typically evaluated by measuring equivalence of the English and the target language dictionaries, which is calculated as the LIWC category frequency correlations in large set of parallel translated texts [6–8]. The translated dictionaries contain more than twice the number of original words [6, 8]. The results show that after machine translation and manual refinement, the resulting dictionaries in target languages reach acceptable equivalence with the English original and can be used as valid measurement methods to analyze texts saturated with psychologically relevant content [6].

The following complications are typically reported in the translation process:

- homonymy, when the translated words in the target languages belong to different semantic categories;
- the wildcards: the asterisk sign (*) meaning that the original word stem should be expanded;
- some translations require finding additional corresponding words in a different culture, not present in the original dictionary;
- specific differences in morphology between English and the target languages.

As a result, there is a risk that direct translation of LIWC, either automatic or manual, does not preserve target language-specific information. The risk increases when translating LIWC into a language belonging to a distant subgroup, such as Russian.

2.2 Russian Thesauri

To address the issues concerning LIWC outlined above, we describe the linguistic structure of the translated LIWC categories in Russian by matching them with two Russian thesauri: **RuThes** and the Synonyms Dictionary (**SynD**).

RuThes. RuThes is a dictionary resource widely used for various NLP tasks [13]. It is a linguistic ontology for Natural Language Processing. The authors stress that RuThes is constructed in terms of ontological, and not lexical relations: it captures relations between concepts, as opposed to linguistic units. The concept-oriented, ontological approach allows for building a highly language-independent dictionary which could be easily matched with information from other languages [14]. Three types of relation between concepts are covered:

- Hyponymy (hyperonymy);
- Part-whole relations;
- External dependence relation, an unsymmetrical relation representing dependence of existence of one concept on the existence of the other.

We use RuThes-lite, a version of the dictionary available for free download. It includes over 111K words and expressions. The terms are organized as concept in a hierarchical structure of hyponymy relations, where each concept is in turn represented by a number of words or multi-word expressions.

Synonyms Dictionary (SynD). The dictionary was compiled by L. Babenko et al. [15]. The dictionary covers 30K words and expressions organized in 5K synonym sets. The dictionary is build based on semantic classification grounds. The words are hierarchically organized in 4 levels:

1. Large semantic domains (15);
2. Semantic classes (84);
3. Semantic groups (255);
4. Semantic subgroups (185).

As **SynD** is built on linguistic grounds and describes the lexical relations, as opposed to ontological relations, it differs significantly from **RuThes** in the following:

- **SynD** is Russian language-specific;
- **SynD** is human-centered:
  - it contains 3.5 times as many words and expressions focused on common human-related domains;
  - its' semantic domains and classes are structured in terms of human life and functioning, as opposed to categories of ontological existence.

In Sect. 3.2, we apply both **RuThes** and **SynD** as hierarchically organized word lists and match the latter with categories of LIWC translated into Russian.

2.3 Author Profiling with LIWC

LIWC is widely used tool in Author Profiling tasks, i.e. related to the language-based personality identification. In fact, LIWC has served as a basis for a recent trend in language-oriented personality modelling [3]. A meta-analysis on predicting individual traits from social media refers to over 30 works based on language, where LIWC usage was predominant [16]. However, the need for changing a paradigm for open-vocabulary approach to Author Profiling is widely discussed [17].

Big Five Personality. In the widely known study by Schwartz et al. [17], LIWC features are used as a baseline for predicting demographics and the Big-Five personality scores. The study is based on a large sample of 75K users, whose Facebook status updates were analyzed in terms of LIWC word category occurrence, and the latter were used in multivariate regression of Big-Five scores. The regressions coefficients reach up to 0.28 in absolute value. In spite of using the large sample and a strict Bonferroni-adjusted correction of significance threshold ($p < 0.001$), most of the obtained results were contradictory to the reference values reported by Yarkoni [18]. The latest work analyses correlations between Big-Five personality scores and LIWC category occurrence in blogs by 694 people. Some unambiguous results confirmed in both works include positive correlations of Biological and Social terms with Extraversion; Social and Time categories with Agreeableness; Biological, Cognitive and Feel categories with Openness. Negative correlations were obtained between Conscientiousness and Bio, Hear, Percept categories.

An early analysis [1, 19] includes LIWC category correlations in essays by 841 participants. A large part of their significant results was contradictory to [17, 18]. This indicates that LIWC category occurrences are not universal and are dependent on the sample quality and size, genre and topic of the analyzed text.

It is important to notice that the absolute value of regression and correlation coefficients between linguistic and trait measures are modest in all the cited works, hardly reaching 0.3 and typically around 0.1.

Emotional Well-Being. A review by Luhmann [20] cites 32 works on well-being analysis in social media data, 11 of which make use of LIWC word categories as prediction features. Most of the studies utilize LIWC emotional categories, *Positive* and *Negative* emotion words. In a famous study Wang et al. [21] suggest a Facebook Gross National Happiness Index to measure the levels of life satisfaction based on Facebook status updates. The Index is operationalized as the amount of positive versus negative LIWC words used in status updates on a given day. Although the Index has not been shown to correlate with life satisfaction, *Negative* words alone are reported to be significantly negatively correlated with it. A number of researchers have since applied emotional LIWC categories in emotional well-being research in Facebook and Twitter texts [22–24]. However, these are often supplemented by pronouns [25, 26], social [27] and causal [28] categories.

Social LIWC categories are found to be contributing to the positive emotion factor, as opposed to work issues and negative emotions [27]. A study on Social Anxiety Disorder reports more *we*-pronouns and less *I*-pronouns in audience speech by participants with the disorder, comparing to healthy ones [25]. Both first-person pronouns (*I, we*) are more often used by depressed individuals, as well as negative emotion, Swear, Anger and Anxiety words in Twitter messages [26]. De Choudhury et al. [29] show that mean values of first-, second- and third-person pronouns usage, as well as social engagement features in Twitter are significant discriminators of depression.

The results reported by different researchers are unanimous in that depression and associated mental health conditions are closely interconnected with usage of pronouns and, more broadly, social words and indicators.

3 LIWC Adaptation into Russian

3.1 LIWC Translation

The following categories were chosen from LIWC [11] for machine translation and analysis as some of the most important for personality profiling: **Cognitive, Social, Biological processes, Perceptual processes** and the contents thereof: **See, Feel, Hear**.

Translation. First, all of the words in each category ending with asterisk (*) were automatically expanded using the English dictionary built-in in NLTK library [30]. In this process, all the words in the dictionary starting with the same stem as the asterisk-stem, but not more than 4 symbols longer than the asterisk-stem, were added as expansions of the stem. Next, each word was translated automatically by using Yandex. API translation service (https://tech.yandex.com/translate/). As single words were translated, there were some morphological errors introduced by the machine translation process, and the translations were lemmatized using PyMorphy [31]. The resulting vocabulary is described in Table 1.

Table 1. Statistics of the translated LIWC categories.

	Bio	Cognitive	Feel	Hear	Percept	See	Social	Time
Words (English)	4,582	2,992	528	666	2,287	487	2,363	1,854
Words (Russian)	1,636	1,251	244	257	903	201	869	795
Words expanded	–	3,199	–	–	–	–	1,678	–

Manual Expansion. Manual analysis revealed that the translated word lists were inconsistent. While preserving some of the concepts present in the original LIWC, other concepts were left out. Words denoting the same concepts in different parts of speech were randomly included in the translated categories. For example, the word *знакомство* ("acquaintance") was included in the **Social** category, however, the respective verb *знакомиться* ("to meet") was not included. Another example is suffixal verb derivation: the words *извиняться* ("to apologize"), *извинение* ("apology") are included in the **Social** category, while *извиниться* ("have apologized"), *извинить* ("have excused"), *извинять* ("excuse") are not.

In order to overcome this issue, the translated categories **Cognitive** and **Social** were manually analyzed by 3 annotators. The annotators were trained linguists. Due to space restrictions, we do not present the full instructions; however, these included the following:

- to delete incorrect translations or words not belonging to the category in question;
- to add relevant synonyms (e.g., *советник* ("advisor"), *консультант* ("*counsellor*")) and words with similar meaning derived from the same root, belonging to the same or different parts of speech (e.g., *сообщник* "accomplice, male" and *сообщница* ("accomplice, female"); *консультация* ("advice"), *консультирование* ("consultancy"), *консультировать* ("to consult").

Inter-annotator agreement was measured with Measuring Agreement on Set-Valued Items (MASI) distance and Krippendorff's alpha [30]. The inter-annotator agreement for both categories remained very low (alpha < 0.2), with distance values very high (~ 0.8). This indicates that manual annotation is not an effective scenario in the vocabulary expansion in our case.

3.2 Matching LIWC and Russian Thesauri

In order to effectively expand the translated vocabulary categories and to overcome the issues of inconsistency and subjectivity, we suggest to supplement the translated categories with existing Russian thesauri.

The translated vocabulary expansion is performed in the following steps:

1. The translated word categories are automatically matched with the categories found in the selected thesauri. The manual expansion information in **Cognitive** and **Social** is preserved.
2. The results of the matching are analyzed manually. The relevant thesauri categories are attached to the LIWC categories, where possible.
3. The expanded Russian LIWC dictionary is constructed based on the selected thesauri categories.

We only include single words in our analysis. Multi-word expressions are out of scope of the current study.

Matching LIWC and RuThes. We have automatically matched the word lists acquired by machine translation (and manual expansion for **Cognitive** and **Social**) with all the categories of **RuThes** (https://www.labinform.ru/pub/ruthes/). We use the hyponymy-based category structure of the thesaurus. The results of the **Cognitive** category matching with both **RuThes** and **SynD** are presented in Fig. 1.

The matched LIWC word categories are represented by a scatterplot, where each category in **RuThes** is situated on a graph: the horizontal axis "InCategory" represents the number of words belonging to the intersection of the **RuThes** category with **Cognitive**; the vertical axis "Ratio" represents the ratio of the **RuThes** category intersecting with **Cognitive**. The size of the circle in the scatterplot represents the number of **RuThes** categories with the same Ratio and InCategory values.

The scatterplots demonstrate that there are only a few **RuThes** categories with at least 10% intersection with the LIWC categories, and the numbers of intersecting words in these typically reaches ~ 20. Moreover, the categories are mostly distributed near to the '0' axes, implying that most categories either intersect with the LIWC categories by a marginal ratio, or contain very few examples themselves. Finally, most categories are situated above or to the right form the (0; 0) point, meaning that most of the **RuThes** categories, however distinct, include a few words belonging to the LIWC categories in question.

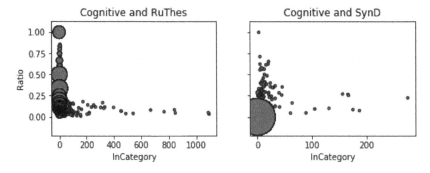

Fig. 1. Volumes of **Cognitive** and **RuThes**, **SynD** categories and their intersections.

The best **RuThes** categories intersecting with **Cognitive** are highly nested: i.e., a number of distinct categories have almost the same intersection with **Cognitive** (*ОПРЕДЕЛИТЬ, ВЫЯСНИТЬ – TO DEFINE, TO DETERMINE, УЗНАТЬ, ПОЛУЧИТЬ СВЕДЕНИЯ – TO DISCOVER, TO FIND OUT, ПОЛУЧИТЬ В РАС-ПОРЯЖЕНИЕ – TO AQUIRE, СООБЩИТЬ, УВЕДОМИТЬ – TO REPORT, TO NOTIFY*), and it is difficult to define, which categories should be chosen on theoretical grounds. These facts indicate that **RuThes** categories are noisy and difficult to analyze in terms of their intersection with **LIWC** categories.

Matching LIWC and Synonym Dictionary. We have performed the same matching procedure for **SynD** categories and LIWC (see Fig. 1).

The results, on the contrary, suggest that most **SynD** categories are situated in the (0; 0) point, meaning that only a few categories actually intersect with LIWC. What is more, for each LIWC category, there is clearly a small number of **SynD** 'outliers' on the graph, situated closer to the right or upper-right side of the scatter plot.

The analysis reveals that there is a small number of clear **SynD** categories intersecting with LIWC categories (see Table 2 for examples). What is more, the structure of **SynD** thesaurus, which is not highly nested, allows for highly informative category names, which can further be used to match with LIWC categories.

Expanding Russian LIWC. For the abovementioned reasons, we have selected **SynD** thesaurus as the resource for expanding LIWC dictionary categories.

Manual examination of the best intersecting categories resulted in the mapping between LIWC and **SynD** categories, described in Table 2.

Table 2 demonstrates that the categories in **SynD** allow for a consistent mapping with the LIWC categories. This is achieved due to three reasons:

1. **SynD** categories are human-centered;
2. the categories are not highly nested (up to 4 levels are allowed in the hierarchy);
3. the categories are clearly and unambiguously described by their names.

The described mapping resulted in Russian LIWC dictionary containing words from the respective **SynD** categories. The LIWC mapped dictionary statistics are described in Table 3. The expanded and disambiguated dictionary contains 9,116 words in normal form. As **SynD** categories contain a lot of ambiguous terms which appear in more than one LIWC category, for the current study we deleted all such words from the dictionary. Table 3 also includes the statistics of the disambiguated dictionary and the presence of the original words from the (expanded) translation.

Table 2. Mapping between LIWC and SynD categories.

LIWC category	SynD categories
Bio	"2.2.8 Structure of a living entity", "3.2. Food and drink as physiological needs of a person", "10.10. Medicine"
Cognitive	"7. Intellect", "15. Universal representations, meanings and relationships"
Feel	"14.6. Tactile perception"
Hear	"14.4. Acoustic perception"
Percept	"14.3. Visual perception", "14.4. Acoustic perception", "14.5. Olfactory perception", "14.6. Tactile perception"
See	"14.3. Visual perception"
Social	"3.1.2. Person of a certain age", "3.1.3. Person of a certain gender", "4.1.6. Friendliness and loneliness", "6. Speech", "11. Social sphere of human life"
Time	"3.1.1. Periods of human life", "10.1.10. Time in education", "14.9. Time"

Table 3. Statistics of SynD categories mapped with LIWC.

	Bio	Cognitive	Feel	Hear	Percept	See	Social	Time
Words (Expanded)	826	3,378	24	432	821	334	4,592	795
Words (Expanded, no ambiguity)	763	2,642	20	303	619	262	3,944	563
Words from orig. translation	82	604	5	11	48	19	340	75

4   Russian LIWC Experiments

We have constructed a preliminary Russian dictionary with LIWC categories by using an available Russian thesaurus and preserving its' category structure. We apply the preliminary dictionary to a number of author profiling tasks in Russian. In line with the original LIWC dictionary, we additionally populate the **Social** category with all the personal and possessive pronouns, except for the first person singular pronouns. Thus the final dictionary applied in this section contains 9,129 normal forms.

We analyze the following author profiling dimensions:

- personality: Big-5 [32];
- well-being: the Hospital Anxiety and Depression Scale (HADS) [33];

We use relative frequencies of the LIWC categories in texts of our subcorpora. Spearman's r is applied to measure the relatedness of profiling measures with LIWC category occurrences. All the experiments are performed with Scipy library [34].

4.1 Dataset

In preliminary Author Profiling experiments with the constructed Russian LIWC we use the datasets from the *RusPersonality* corpus, which is a part of *RusIdiolect* database[1]. *RusPersonality* contains 2,500 texts and extensive annotation. The annotation on text characteristics include mode, genre, topic, deception. The annotation on author characteristics include demographics: gender, age, native language and personality: the Big Five, Freiburg Personality Inventory (FPI), Hospital Anxiety and Depression Scale (HADS), neuropsychological assessment [35, 36].

We used written texts by Russian-speaking authors in our analysis. All the text data in the experiments are natural, implying that the texts are written by lay Russian speakers and are not intended for presentation in blogs or media. The topics of our datasets include the following:

- "What would I do if I had a million dollars?";
- descriptions of 2 different pictures;
- "Letter to a friend";
- "Description of my previous day".

Different datasets in the corpus contain different annotation types. The datasets and annotation used in our experiments are described in Table 4.

Table 4. Datasets in the Author Profiling experiments.

Dataset	Topic	№ of texts	Age: Mean (Std)	Gender: Male (percent)	Text length (tokens): Mean (Std)	Annotation
RusPersonality1	Million dollars	72	22.2 (3.3)	28 (39%)	85.3 (52.8)	Big-5
	Picture	72	22.2 (3.3)	28 (39%)	121.6 (68.6)	
RusNeuropsych	Letter	202	21.2 (3.1)	74 (37%)	192.8 (99.4)	Big-5, HADS
	Picture	190	21.3 (3.0)	68 (36%)	109.9 (60.7)	

4.2 Experiment Results

Big-5 Personality. We have applied the Ruspersonality1 and RusNeuropsych subcorpora to analyze the Russian LIWC category correlations with the Big-5 personality traits. We analyzed the four subcorpora separately, as they belong to different topics, and the goal of the preliminary analysis was to obtain intra-topic LIWC correlations. The significant ($p < 0.05$) positive correlations were obtained for **Extraversion** and **Bio** ($r = 0.24$; Picture in RusPersonality1), **Conscientiousness** and **Social** ($r = 0.14$; Letter in RusNeuropsych). Negative correlations were found for **Agreeableness** and **Hear** ($r = -0.24$; Picture in RusPersonality1), **Conscientiousness** and **Bio** ($r = -0.27$;

---

[1] Freely available for search and download at https://rusidiolect.rusprofilinglab.ru/.

Million dollars in RusPersonality1), **Openness** and **Cognitive** ($r = -0.16$; Picture in RusNeuropsych). Note that in the current preliminary analysis, we do not apply any multiple hypotheses corrections.

HADS Well-Being. We have applied RusNeuropsych subcorpus to identify correlations with the Depression and Anxiety measures. A single significant negative correlation was identified between **Anxiety** and the **Social** category in the Letter subcorpus ($r = -0.2$, $p < 0.01$). The result implies that higher anxiety levels are associated with fewer mentions of words related to social interaction and functioning.

Our results are consistent with previous research both quantitatively (reaching comparable effect sizes) and qualitatively. Namely, the high number of words related to **Bio**logical processes indicating high scores in Extraversion and low scores in Conscientiousness replicates [17, 18]. High Openness is, paradoxically, represented by low numbers of words related to **Cognitive** processes, replicating [18]. Moreover, the negative correlation of words denoting **Social** processes and **Anxiety** is in line with the finding in [27], that **Social** words are related to the positive emotion factor, as opposed to negative emotions and work issues. Indeed, anxiety disorder is often connected to social anxiety, which is sometimes studied as a separate disorder [25].

5   Conclusions and Future Work

We have performed experiments on matching LIWC dictionary categories with Russian thesauri categories.

First, we have automatically translated the following LIWC categories in Russian: **Cognitive**, **Social**, **Bio**logical processes, **Percept**ual processes and the contents thereof: **See**, **Feel**, **Hear**. **Cognitive** and **Social** category translations have been manually enriched by 3 annotators, with very modest inter-annotator agreement. The category translations have been matched with two Russian thesauri categories: **RuThes** and **SynD**. The latter is shown to produce a better match for the LIWC categories and is applied to the development of a preliminary Russian LIWC version, where every LIWC category in question is represented by a number of thesaurus categories. Finally, we have applied the preliminary Russian LIWC to Author Profiling tasks, reaching reasonable results.

Based on the performed experiments, we have arrived at the following conclusions:

1. LIWC categories are not perfectly mapped to the selected Russian thesauri. In fact, the numbers of word intersections are quite modest. **This further raises the issue of subjectivity in LIWC category development.**
2. However, matching machine-translated LIWC categories with **SynD** is a useful way of expanding the translation, which allows to effectively address the issues arising from cultural and structural differences between English and Russian, and overcome subjectivity of LIWC word categories. Thus, **we have constructed a preliminary Russian LIWC dictionary by preserving the original Russian thesaurus category structure.**
3. The selected word categories of **SynD only modestly match the direct LIWC translation** from English to Russian.

4. The suggested preliminary Russian LIWC dictionary can be used in author profiling tasks. Our **results thoroughly replicate the original benchmark findings** on LIWC and Big-Five personality traits.

The modest proportion of words matched between the original translation from English and the selected Russian thesauri categories indicates that the English LIWC dictionary structure cannot be simply reconstructed in Russian for these reasons:

- the vocabulary structure of the English and Russian languages differs significantly;
- original LIWC dictionary content and structure are highly subjective.

This raises an important issue, whether the task of LIWC translation to Russian should be directly pursued. Despite the apparent simplicity of using a resource manually refined by annotators trained in psychology, in practice, the subjectivity issue remains in both vocabulary construction and translation. The validity of the translated vocabulary could be evaluated with English-Russian parallel texts. However, the question of results reproducibility in other genres and modes remains, specifically in analyzing non-professional texts and oral transcripts.

Given these considerations, we suggest to proceed with Russian LIWC development and expansion grounded in existing linguistic thesauri. We propose the following directions for future work:

- inconsistencies between the Russian thesauri categories and the translated LIWC vocabulary should be manually analyzed;
- additional linguistic resources will be suggested, producing a higher recall in matching the translated LIWC categories;
- the rest of LIWC categories, including Affective processes, Drives, Personal concerns, and style categories, should be matched with Russian linguistic resources;
- the resulting vocabulary should be validated with parallel English-Russian texts.

Acknowledgement. The authors acknowledge support of this study by the Russian Science Foundation grant №18-78-10081. The authors are grateful for the considerations provided by the anonymous reviewers.

References

1. Pennebaker, J.W., King, L.A.: Linguistic styles: language use as an individual difference. Trans. Am. Math. Soc. **77**(6), 1296 (1999)
2. Tausczik, Y.R., Pennebaker, J.W.: The psychological meaning of words: LIWC and computerized text analysis methods. J. Lang. Soc. Psychol. **29**(1), 24–54 (2010)
3. Boyd, R.L., Pennebaker, J.W.: Language-based personality: a new approach to personality in a digital world. Curr. Opin. Behav. Sci. **18**, 63–68 (2017)
4. Kailer, A., Chung, C.K.: The Russian LIWC2007 dictionary. LIWC.net, Technical report (2011)
5. Gao, R., Hao, B., Li, H., Gao, Y., Zhu, T.: Developing simplified chinese psychological linguistic analysis dictionary for microblog. In: Imamura, K., Usui, S., Shirao, T., Kasamatsu, T., Schwabe, L., Zhong, N. (eds.) BHI 2013. LNCS (LNAI), vol. 8211, pp. 359–368. Springer, Cham (2013). https://doi.org/10.1007/978-3-319-02753-1_36

6. Bjekić, J., Lazarević, L.B., Živanović, M., Knežević, G.: Psychometric evaluation of the Serbian dictionary for automatic text analysis-LIWCser. Psihologija **47**(1), 5–32 (2014)

7. Van Wissen, L., Boot, P.: An electronic translation of the LIWC Dictionary into Dutch. In: Electronic lexicography in the 21st century: Proceedings of eLex 2017 conference, pp. 703–715. Lexical Computing (2017)

8. Meier, T., et al.: "LIWC auf Deutsch": the development, psychometrics, and introduction of DE-LIWC2015. PsyArXiv (2019)

9. Pennebaker, J.W., Boyd, R.L., Jordan, K., Blackburn, K.: The development and psychometric properties of LIWC2015. The University of Texas at Austin (2015)

10. Litvinova, T., Litvinova, O., Seredin, P.: Dynamics of an idiostyle of a Russian suicidal blogger. In: Proceedings of the Fifth Workshop on Computational Linguistics and Clinical Psychology: From Keyboard to Clinic, pp. 158–167. Association for Computational Linguistics (2018)

11. Litvinova, T., Seredin, P., Litvinova, O., Dankova, T., Zagorovskaya, O.: On the stability of some idiolectal features. In: Karpov, A., Jokisch, O., Potapova, R. (eds.) SPECOM 2018. LNCS (LNAI), vol. 11096, pp. 331–336. Springer, Cham (2018). https://doi.org/10.1007/978-3-319-99579-3_35

12. Pennebaker, J.W.: The secret life of pronouns. New Sci. **211**(2828), 42–45 (2011)

13. Lukashevich, N.V.: Tezaurusy v zadachakh informatsionnogo poiska (Thesauri in Information Retrieval Problems), Moscow, Mosk. Gos. Univ (2011)

14. Loukachevitch, N., Dobrov, B.V.: RuThes linguistic ontology vs. Russian wordnets. In: Proceedings of the Seventh Global Wordnet Conference, pp. 154–162 (2014)

15. Babenko, L.G.: Slovar' sinonimov russkogo yazyka [Dictionary of synonyms of the Russian language]. Astrel, Moscow (2011)

16. Settanni, M., Azucar, D., Marengo, D.: Predicting individual characteristics from digital traces on social media: a meta-analysis. Cyberpsychol. Behav. Soc. Netw. **21**(4), 217–228 (2018)

17. Schwartz, H.A., et al.: Personality, gender, and age in the language of social media: the open-vocabulary approach. PLoS ONE **8**(9), e73791 (2013)

18. Yarkoni, T.: Personality in 100,000 words: a large-scale analysis of personality and word use among bloggers. J. Res. Pers. **44**(3), 363–373 (2010)

19. Mairesse, F., Walker, M.A., Mehl, M.R., Moore, R.K.: Using linguistic cues for the automatic recognition of personality in conversation and text. J. Artif. Intell. Res. **30**, 457–500 (2007)

20. Luhmann, M.: Using big data to study subjective well-being. Curr. Opin. Behav. Sci. **18**, 28–33 (2017)

21. Wang, N., Kosinski, M., Stillwell, D.J., Rust, J.: Can well-being be measured using Facebook status updates? Validation of Facebook's Gross National Happiness Index. Soc. Indic. Res. **115**(1), 483–491 (2014)

22. Settanni, M., Marengo, D.: Sharing feelings online: studying emotional well-being via automated text analysis of Facebook posts. Front. Psychol. **6**, 1045 (2015)

23. Wojcik, S.P., Hovasapian, A., Graham, J., Motyl, M., Ditto, P.H.: Conservatives report, but liberals display, greater happiness. Science **347**(6227), 1243–1246 (2015)

24. Jones, N.M., Wojcik, S.P., Sweeting, J., Silver, R.C.: Tweeting negative emotion: an investigation of Twitter data in the aftermath of violence on college campuses. Psychol. Methods **21**(4), 526 (2016)

25. Hofmann, S.G., Moore, P.M., Gutner, C., Weeks, J.W.: Linguistic correlates of social anxiety disorder. Cogn. Emot. **26**(4), 720–726 (2012)

26. Coppersmith, G., Dredze, M., Harman, C.: Quantifying mental health signals in Twitter. In: Proceedings of the Workshop on Computational Linguistics and Clinical Psychology: From Linguistic Signal to Clinical Reality, pp. 51–60 (2014)
27. Wang, W., Hernandez, I., Newman, D.A., He, J., Bian, J.: Twitter analysis: studying US weekly trends in work stress and emotion. Appl. Psychol. **65**(2), 355–378 (2016)
28. Doré, B., Ort, L., Braverman, O., Ochsner, K.N.: Sadness shifts to anxiety over time and distance from the national tragedy in Newtown, Connecticut. Psychol. Sci. **26**(4), 363–373 (2015)
29. De Choudhury, M., Gamon, M., Counts, S., Horvitz, E.: Predicting depression via social media. In: Seventh International AAAI Conference on Weblogs and Social Media (2013)
30. Bird, S., Klein, E., Loper, E.: Natural Language Processing with Python: Analyzing Text with the Natural Language Toolkit. O'Reilly Media Inc., Sebastopol (2009)
31. Korobov, M.: Morphological analyzer and generator for Russian and Ukrainian languages. In: Khachay, M.Yu., Konstantinova, N., Panchenko, A., Ignatov, D.I., Labunets, V.G. (eds.) AIST 2015. CCIS, vol. 542, pp. 320–332. Springer, Cham (2015). https://doi.org/10.1007/978-3-319-26123-2_31
32. McCrae, R.R., Costa Jr., P.T.: Personality trait structure as a human universal. Am. Psychol. **52**(5), 509 (1997)
33. Snaith, R.P.: The hospital anxiety and depression scale. Health Qual. Life Outcomes **1**(1), 29 (2003)
34. Virtanen, P., et al.: SciPy 1.0: fundamental algorithms for scientific computing in Python. Nat. Methods **17**(3), 261–272 (2020)
35. Litvinova, T., Litvinova, O., Zagorovskaya, O., Seredin, P., Sboev, A., Romanchenko, O.: Ruspersonality: a Russian corpus for authorship profiling and deception detection. In: 2016 International FRUCT Conference on Intelligence, Social Media and Web (ISMW FRUCT), pp. 1–7. IEEE (2016)
36. Litvinova, T., Seredin, P., Litvinova, O., Ryzhkova, E.: Estimating the similarities between texts of right-handed and left-handed males and females. In: Jones, G.J.F., et al. (eds.) CLEF 2017. LNCS, vol. 10456, pp. 119–124. Springer, Cham (2017). https://doi.org/10.1007/978-3-319-65813-1_11

Chinese-Russian Shared Task
on Multi-domain Translation

Valentin Malykh[1(✉)] and Varvara Logacheva[2]

[1] Huawei Noah's Ark Lab, Moscow, Russia
valentin.malykh@huawei.com
[2] Skolkovo Institute of Science and Technology, Moscow, Russia
v.logacheva@skoltech.ru

Abstract. We present the results the first shared task on Machine Translation (MT) from **Chinese into Russian**, which is the only MT competition for this pair of languages to date. The task for participants was to train a general-purpose MT system which performs reasonably well on very diverse text domains and styles without additional fine-tuning. 11 teams participated in the competition, some of the submitted models showed reasonably good performance topping at **19.7 BLEU**.

Keywords: Machine Translation · Shared task · Domain adaptation

1 Introduction

Modern Machine Translation models can perform very well, sometimes almost reaching human performance [5]. However, the high quality can be achieved under certain constraints, namely, large amount of data to train a model and high similarity between the training data and the test data.

There are numerous efforts to mitigate these constraints and make MT more universally applicable. One direction of research is development of models which can produce reasonable translations in low-resource settings: with small amount of data [15] or no parallel data at all [2]. Domain adaptation [3] can be considered a special case of such setting, it is a task of leveraging of large amount of data to translate texts from a different domain with little or no in-domain training data. Another line of work is improvement of MT under particular constraints: translation with pre-set level of politeness [14], simultaneous translation [1], translation of noisy text [10], etc.

Thus, the recent Machine Translation shared tasks usually introduce additional constraints, such as low amount of data, domain shift or additional features, e.g. translation of chat conversations[1] which imply incorporating context beyond a single sentence and modelling of dialogue.

Along this line, our shared task introduces a new constraint for an MT system—the high diversity *within* the test set. Namely, we suggest developing

---

[1] http://statmt.org/wmt20/chat-task.html.

© Springer Nature Switzerland AG 2020
A. Filchenkov et al. (Eds.): AINL 2020, CCIS 1292, pp. 196–201, 2020.
https://doi.org/10.1007/978-3-030-59082-6_15

an MT model which can translate texts belonging to highly diverse domains with no additional fine-tuning. They should be applicable to the setting when sentences from different domains occur in one text and a single MT system has to produce reasonable translations for all of them. This task has been tackled by training on mixed-domain data, combining a bunch of single-domain models [12] or jointly learning to translate and define domain of data [18]. Nevertheless, to the best of our knowledge, there have been no shared tasks on multi-domain MT. The most similar competition to ours is Machine Translation Robustness shared task [8] within WMT conference. However, this task is different to ours, because it explores the ability of a single-domain model to translate data from an unseen domain, rather than the ability to maintain multiple domains.

2   Task Description

We ask participants to train a Machine Translation model for Chinese–Russian language pair. The models are then tested on a testset which contains sentences from two domains: **news texts** and **fiction** in equal shares.

We evaluate the performance of the models with BLEU score [11] in the implementation of `sacreBLEU` package.

The test dataset is compiled of parallel Chinese–Russian news texts and excerpts from fiction books. The overall number of sentences is 6234. This dataset is a part of Russian National Corpus[2] [4]. It includes several short novels of A.P. Chekhov translated to Chinese and news documents from several sources, including Xinhua news agency, translated to Russian.

We did not provide any compulsory training dataset. We suggest participants using the following datasets for the training of models:

- UN Parallel corpus presented in [19]—18M sentences extracted from documents of the United Nations. This corpus is publicly available[3].
- CCMatrix described corpus in [13]—13M web-crawled sentences. The raw corpus CCNet is available, but the filtering criteria which is need to be applied to achieve CCMatrix are not yet released[4].
- OpenSubtitles corpus presented in [9]. It contains 7M sentences extracted from subtitles to TV series and films. This corpus is publicly available[5].
- News Commentary introduced in [16]—47K sentences of political and economic commentary crawled from the web site Project Syndicate[6]. This corpus is also publicly available[7].

In addition to that, we encouraged participants to use any other resources (including pre-trained word embeddings, monolingual datasets and parallel datasets for other languages pairs) as long as they are publicly available.

---

[2] http://www.ruscorpora.ru.
[3] http://opus.nlpl.eu/UNPC-v1.0.php.
[4] https://github.com/facebookresearch/LASER/tree/master/tasks/CCMatrix.
[5] http://opus.nlpl.eu/OpenSubtitles-v2018.php.
[6] https://www.project-syndicate.org/.
[7] http://www.casmacat.eu/corpus/news-commentary.html.

3   Competition Structure

The competition was divided into two phases: *tuning* and *evaluation*. The former allowed participants to compare different approaches, while the latter was the competition *per se*. In order to make the conditions of the tuning phase most closely resemble those of the evaluation phase, we divided our original dataset into two parts (*public* and *private*), one for each of the phases. Neither of the datasets were made available to the participants, they were required to submit their models which were run on our machines.

The tuning phase lasted from April 14th to May 28th 2020. During this period, the participants could submit their solutions and get their scores on the public dataset instantly. We also maintained a public leaderboard, so that the participants could see how their models performed compared to others.

The participants of the competition were required to submit their MT models in the form of docker containers. Each solution had the access to 7 vCPU, 8 GB of RAM and could not connect to the Internet. We set a time limit of 20 min to produce the translation of the whole testset. Each participant was given 5 submits per day during the competition.

The evaluation phase lasted from March 14th to April 30th. The participants were asked to submit their final models of choice, which were run on the private part of the dataset. The BLEU scores of these models on the private testset were considered final results. Each participant could submit at most 5 models per day, summing up to 230 possible models.

4   Results

A total of 11 teams participated in the competition and submitted about 700 systems in total. The results are given in Table 1. The results are also available online on MLBootCamp competition platform[8].

We can see that the first model performs significantly better than all successors, yielding the BLEU of 19.7, whereas the second-best model got only 8.7. This model used Transformer [17] Machine Translation architecture in OpenNMT-py[9] [7] implementation. However, this architecture was used by the majority of participants. Apparently, the key point of the model's high performance is the choice of data. This team extended the training dataset with Russian fiction books along with their translations into Chinese mined from the web.

The majority of participants chose Transformer [17] Machine Translation architecture as the best-performing model for text processing. They used different implementations of Transformer: tensor2tensor[10], Marian[11] [6], and already mentioned OpenNMT-py.

---

[8] https://mlbootcamp.ru/ru/round/26/rating/?results_filter=final.
[9] https://opennmt.net/OpenNMT-py/.
[10] https://github.com/tensorflow/tensor2tensor.
[11] https://marian-nmt.github.io/.

Table 1. Official results of the competition on the hidden test set.

Team/Participant	BLEU
averkij	19.7
firefish	8.7
ualabs	8.6
randomseed19	6.9
kbrodt	6.1
kitsenko	4.6
koziev	1.9
cheremisin	1.7
xammi	0.9
kurochka	0.9
adilism	0.2

Considering the methods of domain adaptation and mixture, most of participating teams followed the path chosen by the best-performing team. Namely, they made their models multi-domain by mixing data of different domains in the training dataset. In addition to the recommended corpora, some participants enhanced their datasets with Russian fiction texts from https://proza.ru/, Russian news texts from Taiga corpus[12] and their automatic translations into Chinese performed by their own models. In addition to that, some participants used Chinese–English datasets with the English part automatically translated into Russian. According to the participants' experience, the most useful parallel datasets from the list of recommended corpora (see Sect. 2) were the UN proceedings and Chinese–English news corpus. *ualabs* team, who took 3rd place, have used trained LASER model[13] described in [2]. Using this model the participant was able to make rough alignment for sentences in translated books available online in Russian and Chinese languages.

Many participants faced the infrastructural difficulties, in particular, the inability to fit their models in the available memory and finish decoding in given time, which did not allow using beam search and increasing the size of models. This suggests that further research should include work on compression of models, which is a major challenge for the task of multi-domain MT, especially when the number of target domains is unknown. Alternatively, we could consider a different competition scenario, namely, ask participants to generate the outputs of the models on their side and only submit the translations. However, for this competition we rejected this setting in order to make all models perform under the same conditions.

---

[12] https://tatianashavrina.github.io/taiga_site/.

[13] The code and the model itself are available here: https://github.com/facebook research/LASER.

5   Conclusions

We conducted the first shared task on Chinese–Russian multi-domain Machine Translation. The specificity of our task (besides the language pair that has not received much attention so far) is the fact that we tested the participating models on texts of diverse domains (fiction and news). The models needed to perform reasonably well on different domains with no fine-tuning.

The vast majority of participants chose Transformer architecture to train the model. The used domain mixing strategies mostly reduced to choosing the appropriate training data. The winning model outperformed the competitors by a large margin due to parallel fiction data included into the training.

In the next editions of the competitions we would like to encourage more sophisticated domain mixing strategies, e.g. by restricting the use of additional data.

Acknowledgements. We are thankful to Kirill Semenov from NRU Higher School of Economics, who generously provided us with test corpus for the competition. We also thank the competition platform provider MLBootCamp (https://mlbootcamp.ru/en/main/) (part of Mail.Ru Group) and Dmitry Sannikov personally for the competition technical support and organization.

References

1. Alinejad, A., Siahbani, M., Sarkar, A.: Prediction improves simultaneous neural machine translation. In: Proceedings of the 2018 Conference on Empirical Methods in Natural Language Processing, Brussels, Belgium, pp. 3022–3027. Association for Computational Linguistics, October–November 2018. https://doi.org/10.18653/v1/D18-1337. https://www.aclweb.org/anthology/D18-1337

2. Artetxe, M., Labaka, G., Agirre, E.: An effective approach to unsupervised machine translation. In: Proceedings of the 57th Annual Meeting of the Association for Computational Linguistics, Florence, Italy, pp. 194–203. Association for Computational Linguistics, July 2019. https://doi.org/10.18653/v1/P19-1019. https://www.aclweb.org/anthology/P19-1019

3. Chu, C., Wang, R.: A survey of domain adaptation for neural machine translation. In: Proceedings of the 27th International Conference on Computational Linguistics, Santa Fe, New Mexico, USA, pp. 1304–1319. Association for Computational Linguistics, August 2018. https://www.aclweb.org/anthology/C18-1111

4. Grishina, E.: Spoken Russian in the Russian national corpus (RNC). In: LREC, pp. 121–124. Citeseer (2006)

5. Hassan Awadalla, H., et al.: Achieving human parity on automatic Chinese to English news translation, March 2018. https://www.microsoft.com/en-us/research/publication/achieving-human-parity-on-automatic-chinese-to-english-news-translation/. arXiv:1803.05567

6. Junczys-Dowmunt, M., et al.: Marian: fast neural machine translation in C++. In: Proceedings of ACL 2018, System Demonstrations, Melbourne, Australia, pp. 116–121. Association for Computational Linguistics, July 2018. http://www.aclweb.org/anthology/P18-4020

7. Klein, G., Kim, Y., Deng, Y., Senellart, J., Rush, A.: OpenNMT: open-source toolkit for neural machine translation. In: Proceedings of ACL 2017, System Demonstrations, Vancouver, Canada, pp. 67–72. Association for Computational Linguistics, July 2017. https://www.aclweb.org/anthology/P17-4012

8. Li, X., Michel, P., et al.: Findings of the first shared task on machine translation robustness. In: Proceedings of the Fourth Conference on Machine Translation (Volume 2: Shared Task Papers, Day 1), Florence, Italy, pp. 91–102. Association for Computational Linguistics, August 2019. http://www.aclweb.org/anthology/W19-5303

9. Lison, P., Tiedemann, J.: Opensubtitles 2016: extracting large parallel corpora from movie and TV subtitles (2016)

10. Michel, P., Neubig, G.: MTNT: a testbed for machine translation of noisy text. In: Proceedings of the 2018 Conference on Empirical Methods in Natural Language Processing, Brussels, Belgium, pp. 543–553. Association for Computational Linguistics, October–November 2018. https://doi.org/10.18653/v1/D18-1050. https://www.aclweb.org/anthology/D18-1050

11. Papineni, K., Roukos, S., Ward, T., Zhu, W.J.: BLEU: a method for automatic evaluation of machine translation. In: Proceedings of the 40th Annual Meeting on Association for Computational Linguistics, pp. 311–318. Association for Computational Linguistics (2002)

12. Sajjad, H., Durrani, N., Dalvi, F., Belinkov, Y., Vogel, S.: Neural machine translation training in a multi-domain scenario. CoRR abs/1708.08712 (2017). http://arxiv.org/abs/1708.08712

13. Schwenk, H., Wenzek, G., Edunov, S., Grave, E., Joulin, A.: CCMatrix: mining billions of high-quality parallel sentences on the web. arXiv preprint arXiv:1911.04944 (2019)

14. Sennrich, R., Haddow, B., Birch, A.: Controlling politeness in neural machine translation via side constraints. In: Proceedings of the 2016 Conference of the North American Chapter of the Association for Computational Linguistics: Human Language Technologies, San Diego, California, pp. 35–40. Association for Computational Linguistics, June 2016. https://doi.org/10.18653/v1/N16-1005. https://www.aclweb.org/anthology/N16-1005

15. Sennrich, R., Zhang, B.: Revisiting low-resource neural machine translation: a case study. In: Proceedings of the 57th Annual Meeting of the Association for Computational Linguistics, Florence, Italy, pp. 211–221. Association for Computational Linguistics, July 2019. https://doi.org/10.18653/v1/P19-1021. https://www.aclweb.org/anthology/P19-1021

16. Tiedemann, J.: Parallel data, tools and interfaces in OPUS. In: Proceedings of the Eighth International Conference on Language Resources and Evaluation (LREC 2012), pp. 2214–2218 (2012)

17. Vaswani, A., et al.: Attention is all you need. In: Guyon, I., et al. (eds.) Advances in Neural Information Processing Systems 30, pp. 5998–6008. Curran Associates, Inc. (2017). http://papers.nips.cc/paper/7181-attention-is-all-you-need.pdf

18. Zeng, J., et al.: Multi-domain neural machine translation with word-level domain context discrimination. In: Proceedings of the 2018 Conference on Empirical Methods in Natural Language Processing, Brussels, Belgium, pp. 447–457. Association for Computational Linguistics, October–November 2018. https://doi.org/10.18653/v1/D18-1041. https://www.aclweb.org/anthology/D18-1041

19. Ziemski, M., Junczys-Dowmunt, M., Pouliquen, B.: The united nations parallel corpus v1.0. In: Proceedings of the Tenth International Conference on Language Resources and Evaluation (LREC 2016), pp. 3530–3534 (2016)

Author Index

Printed in the United States
By Bookmasters